WHY ME?

T0348955

WHY ME?

MY LIFETIME OF ALIEN ENCOUNTERS

A memoir by

ROBERT HUNT

As told to Dee Willson

SUTHERLAND
HOUSE

TORONTO, 2024

Sutherland House
416 Moore Ave., Suite 304
Toronto, ON M4G 1C9

First edition, July 2024

If you are interested in inviting one of our authors to a live event or media appearance, please contact sranasinghe@sutherlandhousebooks.com and visit our website at sutherlandhousebooks.com for more information.

We acknowledge the support of the Government of Canada.

This book contains information obtained from authentic and highly regarded sources. The author does not assume responsibility for the validity of materials or the consequences of use.

Manufactured in Canada
Cover designed by Vera Lluch & Lena Yang
Editorial assistance from Denise Willson, Beop Inc., www.beop.ca
Book composed by Karl Hunt

Library and Archives Canada Cataloguing in Publication
Title: Why me? : my lifetime of alien encounters / a memoir by Robert Hunt.
Names: Hunt, Robert (Alien abduction victim), author.
Identifiers: Canadiana (print) 20240297717 | Canadiana (ebook) 20240335058 |
ISBN 9781990823589 (softcover) | ISBN 9781990823848 (EPUB)
Subjects: LCSH: Hunt, Robert (Alien abduction victim) | LCSH: Human-alien encounters. | LCSH: Alien abduction. | LCGFT: Autobiographies.
Classification: LCC BF2050 .H86 2024 | DDC 001.942092—dc23

ISBN 978-1-990823-58-9
eBook ISBN 978-1-990823-84-8

For those who have been visited or abducted by aliens.
I understand. And you are not alone.

CONTENTS

INTRODUCTION

My name is Robert Hunt and I've been abducted by aliens my entire life. This is no prank. I am not creative enough to make shit up.

I am also not crazy, delusional, or mentally unstable. I am intelligent, educated, and a master electrician within the nuclear industry, a job that requires physical and situational awareness at all times. While I have endured the odd moment of despair when life throws sharp objects, I do not struggle with depression. I have never attempted to end my life—I happen to like my life, even the hellacious parts (Did I mention six decades of alien abductions?). I am not, nor have I ever been, an addict. I have no interest in gambling, drugs, or the excessive use of alcohol. Sure, I've enjoyed a beer or two with friends and family and have made my share of toasts on special occasions, but I have never fought the demons of alcohol or drug misuse. I could even give up chocolate if compelled. Most who know me would say I'm a positive guy. I respect those around me and aim to see the best in people. I would never hurt anyone. I can't even stomach violence on television. I like puppies *and* kittens. I give money to the poor. I've been happily married for twenty-nine years. My wife, Lisa, is a lawyer and a partner at a respected Toronto law firm. She doesn't take any crap from me, which is one of the many things I love about her. We are the proud parents of two children. Our daughter, Keira, is a senior publicist for a public relations firm. Our son, Josh, is studying biochemistry with plans to run a lab one day. We love to ski and travel. We are book people, an educated bunch.

This book is not a work of fiction. It is a record of my life, of sixty-three years as a human being who has seen and experienced the unexplained, the unfathomable, the impossible—and lived to talk about it.

I have no delusions of grandeur. I am not interested in the spotlight or stardom. My experiences are hard to swallow, even for me. They challenge politics, religion, science, and reality as we know them. I have no doubt the critics and skeptics will swarm. They're priming their verbal slingshots, readying for the kill. There was a time I feared the vultures. For most of my life, I've worried about what people would say if they knew the truth—my gut has churned over my career, my family, my relationships, and my reputation. I *know* why abductees don't come forward. But I am not afraid anymore. Time, experience, and age will do this to a man.

I don't need others to point out the holes in my life. I do that just fine on my own. Not a day goes by that I don't question what I've experienced. I don't, for one minute, think I am fully capable of truly understanding the depth of what I've seen, felt, and learned. I have more questions than answers. And I am the king of denial.

I wrote this book for the curious, the open-minded, and those who suspect there is *more*. I'm also sharing my story for the others—for the souls who have been ripped from their beds or cars or wherever their feet were planted by creatures few believe exist. I've endured your terror. I've experienced the wonder. I understand the burden you carry every single day, every hour, every minute, and every second. I know the weight is too much to bear. And while I can't take away the pain, confusion, or feelings of isolation, I can assure you, you are not alone.

We are never alone.

1

MAY 26, 1964

FIRST ENCOUNTER

My first encounter with extraterrestrials took place at 93 Beaconsfield Avenue, London, Ontario. It was the night of May 26, 1964. I remember it vividly. It was my seventh birthday.

Earlier that day, my mother had given me Tammy Ten wallpaper as a surprise gift. Tammy Ten was my favorite race car and I was excited to help my parents paper my bedroom. I hoped that the new look would make the old house, and my small room, about eight feet wide by twelve feet, right next to the basement stairs, feel a lot less creepy. When you entered my bedroom from the kitchen, the bed was to your right, and the window was straight ahead with the closet to the left.

I have no idea what time it was, but I remember waking to the sound of my closet door opening. It made a squeaking noise and I was instantly alert. I watched, paralyzed with fear, as the door opened all the way. I noticed a ringing in my ears, a pressure that felt all-consuming. At first, I couldn't see anything. The room was dark except for a sliver of light shining in from the service station next door. I saw the outline of something half my size moving across the floor, backlit from the service station light. I knew it wasn't my imagination.

I tried to scream for my life but only a choking sound came from my throat. I remember knowing no one could hear me, although I didn't know how or why. That made me panic even more.

The dark shadows of three small figures moved toward me. My body was shaking uncontrollably. I thought it was the bed shaking at first, until I noticed it was me. I didn't feel pain—at least not that I remember—but the movement frightened me greatly. I knew, although I'm not sure how, that these figures were not human. I felt pinned to the bed, unable to run from the room. *Someone help me!*

One creature came right up to my bed. It wasn't any bigger than the dog down the street, but it walked upright and as best as I could tell in the dark, it had what looked like a head and two arms. I tried to scream again and managed to pull the bedsheets over my head. A voice said, "Don't be afraid. I am your mother." I don't think I realized it at the time, but the voice came through my mind. It was a calming, female voice, and it paused my sobbing. "Don't be afraid," she said again, gently tugging the sheets from my hands, "I am your mother."

I don't know why, but I trusted her. I knew this wasn't my mother—obviously—but something felt right. I felt a weird sensation take over, a feeling I couldn't put into words. My body was still vibrating, but I wasn't as afraid. I let the creature lower the bedsheets a bit.

I'm not entirely sure what happened next, but I remember her speaking to me in that soothing voice, repeating, "Don't be afraid," while running what I assumed were her fingers through my hair, over and over, until a darkness took over and I fell asleep or passed out.

That's all I remember from my first encounter.

The next morning, I sat at the breakfast table with my mother and three sisters and tried to tell them what had happened. My sisters were unsympathetic. They made ruthless fun of me, saying I was being haunted by ghosts from the basement. I was the second youngest, each of us less than a year apart. As the only boy, I was always outnumbered. "They're gonna eat your brains while you sleep," Wanda said. Sharon and Linda added the visual effects while my mother laughed.

I insisted on the truth of my story. I gave them details that should have been irrefutable. I couldn't make a dent in my family's convictions. Tears welled in my eyes and spilled down my cheeks. No one believed me.

I didn't give up. For days, I did my best to convince them and begged for their help. *What if these things could eat my brains?* The thought of the creatures returning made my stomach hurt. I got nowhere.

Rather, my mother's patience began to run thin. This was not acceptable talk, she said: "Stop concocting tall tales. It was a bad dream, nothing more."

For a long time, I believed her.

2

1964

NIGHTMARES

The nightmares didn't subside. They got worse. After the night of my seventh birthday, I was visited by the creatures over and over again. "My mother" and the other two removed my bedsheets, stripped me naked, and forced me from my bed. Time is a loose concept when you're seven years old. Thinking back to the months and years after the first encounter, I'm not sure if the visits were a weekly or a monthly occurrence. But I did learn to keep them to myself. My actual mother was a foundry worker who made door handles and horn buttons for General Motors and Ford. She wasn't the soft sort. If I uttered a word about creatures or otherworldly beings, she would get upset and pull out the belt. She knew how to use it. When she'd had enough of what she called my lies, she'd called my father. His use of the belt shut me up good.

Despite this, I have a difficult time looking back on those early visits with negativity. A lot has happened in my sixty-something years of alien contact, and I'm still here to talk about it. That said, the word nightmare still comes to mind when I think back on those earlier abductions. I was so very young.

It wasn't only the creatures or the spaceships or being stolen from my bed that frightened me. It was the glass rod.

I knew the creatures were in my bedroom before I could see them. I could feel them. The anticipation was brutal. My body would shake, a vibration that woke me at all hours of the night. The sensation wasn't painful, but I fought it until a blue-green light engulfed the room. At first, I was afraid to open my eyes, but when I could feel the creatures removing my pajamas, I reacted as anyone would—I tried to stop them. Between them, they had three sets of hands. No matter how many times I pushed them away, they were always faster.

The room was dark, but I could make out their shape and size in the blue-green mist. Like humans, they had a torso with arms and legs topped with a proportioned head. But that's where the similarities ended. They were covered in what looked like light brown animal fur, except for the palms of their hands and faces, which were bare. If there was fur on their face, it was too short to see. Small ears stuck out from either side of their head, and a tuft of longer, darker-colored hair topped their heads. Their noses were positioned like ours, but they protruded differently, the shape reminding me of a soda can. There was a dark pattern across their nose, as if they wore black sunglasses. Their eyes were big and solid black, making it hard to know where they were looking. They didn't use familiar body language or facial expressions I could decipher.

The creatures didn't hide or avoid eye contact. They didn't seem to care if I saw them at all. While they removed my bedsheets and clothing, they would talk to me in words I understood—in English—through mouths that were nothing but Hot Wheel-length slits that never opened or moved.

If they had a scent, I couldn't detect it. If they moved in an unfamiliar way, I don't recall how to describe it. If their touch left me with an explainable sensation, I don't know how it felt. These are not details I remember. The year was 1964. American girls were going bonkers for the Beatles, and everyone was watching *Bonanza* and *Bewitched* on television. The Star Wars movies wouldn't be made for almost two decades. Yet, if these creatures resembled anything I've ever seen, I'd say they looked like the Ewoks from *Return of the Jedi*, released in 1983.

The same three beings came every time, and I quickly came to recognize the one who referred to herself as "my mother." I was seven, somewhat tall and lanky, and the top of her head reached my chest. It didn't seem that there were physical differences attributable to gender, but their voices were distinct, clearly male or female, so I always recognized her voice. She had a way of calming me, a way of making me feel at ease when I knew the situation called for unabashed panic.

Once my pajamas were removed, my body was lifted into the air, floating up through the ceiling. I had no sensation of movement—it felt no different than lying on my bed. But I could see I was moving upward. The creatures were with me, floating while standing upright. One always held my hand. Understanding what I saw was difficult. I could see through the ceiling and attic as if the blue-green light dissolved everything it touched. I don't remember being afraid. If anything, I was fascinated, especially when I saw where the light came from, the smooth underside of something big floating in the sky.

I suspect alien ships or spacecraft didn't come to mind in those days, but I can't be certain: I'm not sure I knew what those were. And time makes some memories muddy.

I'm sure I didn't see any doors or hatches or beings pull me inside the spaceship. It was hard to see through the blue-green light that surrounded me. Also, what I saw wasn't always the same—the same *whatever it was*. The floating objects came in different shapes and sizes. I do know that's when I passed out—every time.

I always woke in a strange room that looked nothing like any room I'd ever been in or seen on television. The walls were rounded, shiny, and bright white. There were no visible doors or windows or light sources. Sometimes the floors were white, sometimes gray, and sometimes the floors were lit like the walls. Scattered about were pieces of equipment, machines that reminded me of a hospital, only they didn't make a sound.

I would be on a stretcher or examination table of some sort, with at least two of the creatures in the room with me. The tables were hard and cold. I was never tied down or pinned in any way. They'd prop me up so my legs were dangling over the edge. My feet would almost touch the floor, but I never once thought to jump off and run. I have no idea why. Maybe they kept me docile. Perhaps I instinctively knew there was nowhere to go.

I should have tried.

One creature would hold me upright, tilting my head back, and the other would stand in front of me with that glass rod. It resembled an icicle, about the length of my mother's knitting needles. The creature held the thicker end, and the tapered end was rounded, not flat or sharp. It looked like it was made of glass. I panicked every time I saw this rod, and a male voice would try to calm me, telling me not to be afraid. "This won't hurt," he'd say on repeat. "This goes beside your eye. You will not feel it."

I know I screamed—or tried to scream. I remember trying to move, to push the creature away every time he came toward me with the glass rod, but I couldn't stop it from happening. I had no control over my movements. I could only watch as the creature slid the glass rod into the socket of my eye, beside the bridge of my nose.

I could feel the rod slowly going in. It was cold, but there was no pain. Whatever this procedure was, it didn't hurt, though I knew it should. The sight of a glass rod going into my head was disturbing. Once inside, the rod seemed to illuminate my vision—all I could see was a blinding white light with strange auras of red and blue.

I would sit super still, petrified.

A short time would pass—seconds or possibly minutes—until the rod was slowly removed, the colors and bright light going with it. There was no blood, no mess. No calming words were spoken—no bandages offered. I don't remember anything more than the creatures laying me down on the stretcher when the procedure was complete.

Sometimes, I'd wake in my bed back home as if nothing had ever happened.

Other times I'd wake in another all-white room, a place I came to think of as school. The schoolroom looked the same as the operating room, only there was a chair and a small white table with rounded corners instead of the machines. The furniture was unlike anything I'd seen before. The table felt like plastic and was hard to the touch. But the room itself was what interested me. It had no depth. There didn't seem to be walls or a ceiling, just light without a source. If I was brought in through a door, I couldn't see it. If there were sounds within this room, I couldn't hear them.

The place felt like absolute nothing.

I could only hear the voice of the creature I'd come to know as my mother. I couldn't see her. But she'd talk to me, calmly at first, saying ordinary things, like, "Hello, Bobby, it's nice to see you again." And I'd sit at the table.

Whatever happened in this classroom, over and over, I can't recall it now. These memories are surrounded by fog. I remember her asking questions I couldn't answer, cowering when she raised her voice, and a pressure in my ears, an ache that hurt. I'd cover my ears with my hands, but the pain didn't leave.

"Come now, Bobby," she'd say. "You are fine." The pressure would increase if I didn't respond.

Decades later, I'm unable to piece together those lessons. I only know they were meant to teach me something. I was the student.

3

NEAR-DEATH EXPERIENCE

In July of that same year, nearly two months into my alien abductions, I had another experience that forever marked my memory.

It was a sweltering Sunday afternoon in late July. The sun had a pulse, and there wasn't a single cloud in the piercing blue sky. We were at my Aunt Marion's place, where we often went on weekends when the weather got hot. My father's sister and her husband, Uncle Keith, had a pool with a slide and diving board, and my sisters and cousins and I would spend the entire day staying cool in the pool.

I was a skinny kid. My hair was very blond, and my dad would shave my head with brush cuts I would come to hate in a few years. Our parents sat on the deck that protruded from the back of the house talking and drinking beer from cans pulled from the cooler. Every once in a while, they'd yell at us to quiet down.

On this particular day, we were grating on the grownups' nerves; so, to keep us occupied, Uncle Keith gathered us at the pool's deep end to play a game. It was simple, really. He called it "Fetch." He reached into his pocket, pulled out a fistful of change, and tossed the lot into the deep end of the pool. "You get to keep all the money you collect as long as you take turns diving in for it," he told us. "The ice cream truck is due to come around any time now." He looked right at me. "Take turns, no fighting, and stop bugging your parents."

Off we dove, taking turns, trying to see who could come up with the most change in one breath. The adults got what they wanted; we were having a blast, not paying them any attention. While waiting for our turn, we listened intently for the ice cream truck's bells. I didn't notice Uncle Ron, the family joker, climb onto the roof of the pump house. The pump house was a shed beside the pool that housed the pool pump, with a small area where we used to get changed. It wasn't meant for climbing on or jumping from, but Uncle Ron had had a bit too much to drink. I never saw him launch himself from the roof. I was at the bottom of the pool collecting change that kept falling through my fingers.

Almost out of air, I turned and started toward the surface when a loud splash broke the water above me. As Uncle Ron's large body passed mine, his momentum pulled me back down to the bottom of the pool.

I had to breathe, so I took a massive breath.

I remember the cold water rushing through my lungs and an instant of panic before a sense of overwhelming calm came over me. I wasn't afraid, didn't struggle. I simply stared at the blueness surrounding me, not knowing or caring if it was the blue cement of the pool or the cloudless sky. The coins between my fingers caught the sun in a blinding light.

Then everything went black.

The next thing I knew, I was hovering above the pool, facing the shallow end. I could see the water below me, several feet down. The water was still. The pool was empty. I had no body, yet I was floating. My absence of a body didn't bother me. I felt completely content.

I could see a large group of kids and adults huddled on the walkway beside the pool stairs. A man was kneeling, covering a child's body; only scrawny legs protruded from under the man. I couldn't see what the man was doing, but I could hear him saying, "Come on," repeatedly. Coins were scattered on the concrete. I could see and hear everything that was happening, but I didn't recognize anyone. This was me, only not. I had no clue who these people were or who the child was. I didn't even know if the child was a boy or a girl. But I watched as the people cried, understanding they were afraid. And I knew this child had drowned.

I felt no pain. I didn't see a tunnel or bright light. My life wasn't flashing before my eyes. No deceased family member beckoned or joined me. It was just me, and I was okay. I didn't wonder how I was floating above the pool or search for a body I didn't have.

I have no idea how much time passed before a sudden wave of pain engulfed me. I had a body again, and it hurt. A man's voice—strong and clear—comforted me, saying words I didn't necessarily understand but found soothing nonetheless.

Now, as an adult, I reflect on this experience and try to fill the gaps in logic. I'm told several minutes passed before anyone noticed I was not coming up from the bottom of the pool. Kamikaze Uncle Ron had already climbed out. Then my mother screamed and Uncle Keith dove in to get me. I wasn't breathing when he splayed me on the ground. My lips had turned blue. I reckon I died in that pool, or came damn near close to dying. I believe the voice I heard was comforting me during the transition back to my seven-year-old body. I think the pain was me experiencing a reconnection of nerves and tissue and experiencing the effects of drowning.

When I opened my eyes on that sweltering day in July 1964, coughing and vomiting in the body of my youth, the experience had left a permanent mark on my memory. There was no lasting physical damage, but it raised questions I'm still asking myself.

Had the glass rod procedure changed me in some way? Do we experience reincarnation? Is there a link between near-death experiences and celestial encounters? Did I learn a lesson I was meant to know? Is a part of us immortal? Do we have souls who carry on?

Even now, I often consider something the voice—whoever or whatever he was—said to me that day. "It is just the pain of life," he said rather nonchalantly. I don't know why I found that comforting, but I did. I still do. It is the one thing I understood at the time. The pain I was experiencing was part of living, of life. It was not only necessary but significant, and I shouldn't be afraid of it. I should embrace it.

Since that hot summer day when I watched myself outside myself, I haven't feared death. I want to live, to thrive, but the end does not scare me.

Perhaps this was my lesson to learn.

4

VALIDATION

The twelve years after my seventh birthday passed in a blur of alien abductions, procedures, and lessons, alongside more typical pubescent experiences. As birthday after birthday went by, the somewhat irregular nighttime occurrences became a routine part of my life. It was always the same Ewok-like creatures back then, and I don't recall feeling like they wanted to hurt me. Still, at some point in those early years, my body's self-defense mechanisms took over, and when the beings started to remove my bedsheets or clothing, my mind checked out. The whole experience was just too much to deal with. I was trying my best to ignore it. *Nothing's happening. Mouth shut. End of story.*

I grew facial hair, liked girls, and learned to sleep well when I knew the creatures weren't coming. I could always sense when they *were* coming. I'm not sure if it was some sort of internal clock or knowing a certain amount of time had passed, but an anxious feeling rose from my belly when I knew the creatures were due to visit. It was as if I could feel their vibration before they were in my room. On those nights, I'd have trouble sleeping.

In May 1976, I turned nineteen. I didn't know it at the time, but my life was about to change dramatically.

It was a dry, cool summer, perfect for a nineteen-year-old with a motorcycle. My buddies and I liked to ride late at night when the streets were ours.

I remember it was late August. The night air was crisp with the faint taste of fall, and my friends were in a frenzy. Summer was coming to an end, and we were taking great pains to avoid the back-to-school blues. I was heading into my first term at Fanshawe College for an electrical apprenticeship. This wasn't going to be an easy year, but I desperately wanted to be a master electrician.

I recall that night clearly, although it is shrouded in doubt. Or maybe it was beer goggles. I lived in blue jeans, t-shirts, and a leather jacket— garb that made me look like a hoodlum, according to my mother. My buddies and I had been causing a ruckus at a tavern on the opposite side of town. We liked being single and reckless. When we called it quits, there were four of us hooting and hollering, relishing in the drive across the city on our motorcycles. Dawn was threatening to overcome the darkness by the time I got home, and I was well aware I'd had too much to drink.

I was living with my parents and my younger sister, Linda, at the time, and my mother wasn't fond of my nocturnal tendencies. It would be decades before I had teenagers of my own and came to understand parental worries.

I walked on eggshells when entering the house, trying to ghost myself to my bedroom, which was across the hall from my parents' room. I was relieved when I finally made it to my room and locked the door behind me. Like most teenagers, I liked to sleep in until noon and preferred to do so without explanation.

Soon after I crashed, I was startled awake by a blinding light.

I sat upright in bed, scooting my back against the headboard, and covered my eyes with both hands. The light was so bright I couldn't focus. At first, I thought the house was on fire. The light engulfed the entire door of my bedroom. It scared the crap out of me.

The fear passed in a flash. There was no heat, no smoke, and no sound.

As I struggled with the light, and to understand what was happening, I saw the complete outline of a person standing by my doorway. Confusion hit me hard. I was used to seeing interstellar beings, but this looked like a human man surrounded by the sun. His arms were stretched toward me, wide, as if he was embracing the sun itself. The man seemed huge, a lot bigger than the door. He had what looked like long hair, but I couldn't see much more than his silhouette. He spoke—of that, I'm sure. But I have no idea what he said to me. I'm not sure if I didn't hear or simply can't remember. I was too freaked out.

The entire episode was over in minutes. When the light and man disappeared, I crumpled in bed. *What the hell just happened?* This was not a dream or a drunken hallucination. Sure, I was tired and buzzed, but I wasn't asleep. I was wide awake. My senses were on overdrive. I was shaking, frightened, and maybe even in shock. If not for the effects of the alcohol from the bar earlier, I doubt sleep would have come at all that night. But it eventually did.

I wish I could remember more of this encounter. I wish I understood who this person or thing was and why he'd come to my room.

What I do know is this was the first encounter I didn't experience alone.

The following morning, I woke to find the bed completely soiled. We had this little white yappy thing that would piss on the floor sometimes, and my first thought was to blame it. I threw on jeans and a dirty t-shirt and rushed to the door, set to find the bugger, but my bedroom door was locked from the inside. I'd locked it before bed. The dog wasn't at fault. I'd pissed myself.

I stumbled to the kitchen where my father was sitting in his rocker in front of the bay window, drinking coffee. My dad was a trucker. He drove big rigs and worked long hours alone, making him a man of few words and many cigarettes. He wore no shirt, just jeans. As usual, he had a smoke hanging from his lips and his shoebox of homemade cigarettes open on the table. He was cutting an apple into pieces to tuck inside the box. He said they kept his cigarettes fresh.

I was nervous, hoping he hadn't seen anything in the night. My dad was a no-shit kind of guy, and I preferred not to piss him off. I poured myself a coffee and sat in the chair next to him.

"How was your night?" he asked.

My heart beat faster. "Same as always," I said.

So many words fought their way to my tongue, but I couldn't get them out. I'd long since understood he had no patience for alien talk. My father was a nonbeliever in all things you could not prove beyond a shadow of a doubt. Speculation wasn't in his vocabulary. Curiosity was for silly girls. He thought apparitions and signs were things crazy people muttered.

My father stood to leave the kitchen, then paused and turned to me. "What the hell were you doing last night?" he asked. "Did you have some kind of floodlight on in your room?"

Shit, he saw the light!

It was my chance to spill everything. I wanted to tell him about the man surrounded by the light, to tell him aliens were real. I wanted to explain the years of waking to vibrations and magic blue-green light and to describe the creatures who abducted me. I wanted him to know about the lessons and the glass rod. I wanted him to hear I wasn't afraid of dying and that I was pretty sure we have souls that carry on. Most of all, I wanted him to believe me.

"No," I said instead. "Only my bedroom light was on. Then I went to sleep."

"Strange," he said. "Your ma and me saw a bright light around the doorway. We thought the damn house was on fire. I jumped out of bed and put my hands against the door, but there was no heat. Even the door handle was cold."

I covered my mouth, unable to speak.

He squinted, looking me over. "You must have been doing something in there—that wasn't your bedroom light. Your mother could see dust particles in the air."

For once, my father wasn't the only man of few words. I just stared into my coffee cup.

"When I opened our bedroom door," he said, "nothing was there. The hallway was completely dark. I knocked on your door, but you didn't answer." He looked torn, like he didn't know whether to pat me on the back or crack me one. "I tried to check on you," he added before leaving the room, "but your door was locked."

My breath escaped my lips in a rush. Forty-one years would pass before this night made sense to me.

5

GHOST

My life was never the same after the night I saw the man surrounded by light. Nothing seemed to change for my parents—they were still rock-solid nonbelievers I couldn't confide in—but my perspective had changed. I was on the cusp of manhood, no longer that adolescent boy who needed to bury strange experiences in the dark regions of my subconscious. The proverbial floodgates were opening.

I wanted to know more about these encounters. I'd long ago learned the average kid wasn't abducted by aliens on a regular basis. I knew no one talked about it if they were. Still, I felt the need to validate my experiences, to convince myself I wasn't crazy or delusional. I'd experienced things beyond words, beyond reality, yet I had no doubt they were real.

Above all, I wanted answers to the questions that plagued my thoughts. Questions like: *Why me?* Keep in mind, this was 1976. Cell phones, home computers, and the worldwide web were the domain of sci-fi movies and people with overactive imaginations. If you wanted to know something, you went to your family, teachers, or the library. If you were really lucky, your parents had been swindled by a door-to-door salesman and the bottom three shelves in your living room held vestige to a dusty set of tombs referred to as "the books" or "the encyclopedias."

As you can imagine, none of my questions about extraterrestrial experiences could be answered through an available outlet.

I was only four when Barney and Betty Hill were abducted. At some point, I'd heard of the Hills, the American couple who were taken while driving through rural New Hampshire in 1961. Who hadn't? As the first widely publicized report of alien abduction in the Western hemisphere, their story spread through the media and a best-selling book was published in 1966.

The couple claimed a pancake-like saucer flying every which way in the night sky hovered over their car before Barney slammed on the brakes in the middle of the highway. The couple was taken aboard the UFO, where aliens performed tests on their bodies before returning them to their car. But in 1976, fifteen years had passed since the Hill event and the details that came my way were nothing more than rumors riddled with laughter. Their case was of limited value to me. At nineteen, I was on the verge of freeing my childhood traumas, but I couldn't talk to anyone—who would believe me? Frustration ruled, and the longer my questions went unanswered, the more I needed to push for them.

I was still at Fanshawe College, studying for my electrical license, and working part-time at Canadian Tire where I stocked shelves and helped customers in the hardware department. I was hardworking and friendly—customers liked me. I still lived with my parents and Linda. My older sisters, Wanda and Sharon, had moved out. Sue, my girlfriend, lived about twenty minutes across town.

The Ewok-like creatures were still in my life but on a much smaller scale. I hadn't seen them in months. They offered no send-off or fanfare; they just faded into the background like the end of an old black-and-white film. So much was happening, changing, and I was racing to keep up with it all. Between school and work, I was exhausted. I'd developed asthma and hadn't felt well for months. When I wasn't with my girlfriend, I usually hit the pillow early, before eleven o'clock.

For weeks, I'd been having the same nightmare, like a song on repeat.

I was a small boy, maybe four or five years old. If I was actually me at that age, I didn't recognize myself. I just knew I was a young boy. Something urged me to crawl under my bed, where I was then swallowed by a black hole. This was no *Alice in Wonderland*. Frightened, I flailed about, falling to my death. The darkness was all-encompassing. I could feel a damp, cold wind rushing past. My stomach hurt, bunched into knots. I had no idea when I would hit bottom or what would happen to me when I did, but I was terrified to find out.

At some point, a man's voice echoed through the abyss. "Only you can control your fall," he said. But that made no sense to me, I hit the bottom of the darkness and woke in a sweat, my bed showing obvious signs of distress.

While I knew this was a nightmare, I was still afraid to sleep. The same thing would happen, night after night, and the same voice would tell me only I could control the fall. While this sounded like a great idea, I had no clue what it meant or how to stop myself from falling. I only knew I had to get the nightmare to end.

On the last night I had this nightmare, I screamed out loud at the top of my lungs. "Stop falling!" I was angry. I was tired and I'd had enough. To my surprise, everything stopped. The black hole still existed, but I was no longer descending through the darkness. After a few seconds, the blackness faded away, and I woke in my bed.

Something had changed. I felt it, but didn't understand the feeling.

A few nights later, I opened my eyes to find myself floating above my bed, with no extraterrestrials within sight. There was no blue-green light. I couldn't sense an otherworldly presence, and my body wasn't shaking or vibrating. In fact, I didn't feel much of anything. There I was, looking down upon myself, staring at my body only three or four feet below in my bed. The me in the bed was sleeping on my side with my head askew to the right. The sheets were bunched around my knees. I could see my chest rising and falling in a deep sleep. *How could this be?* I had no idea how it happened, but I was sure it was me sleeping in my bed and also me floating above.

The feeling was incredible.

This was not like the near-death experience I had when I was seven. This time, I felt connected to my body. I could think, see, and move around knowingly. I was aware of who I was, where I was, and what I could do. I felt like myself—better than myself, like Robert 2.0.

I had an overwhelming sense of freedom. Curiosity thundered through my veins like a locomotive. I was simultaneously excited and scared, and I wasn't worried about straying too far from my body. My gut said I would be fine. I left the bedroom and didn't give my body a second thought.

My parents were sleeping in the room across from mine. The house lights were off but I could see. I considered peeking in on my sister but decided that was a bad idea. I spun in circles, taking everything in. I noticed my movement was driven by thought. When I wanted to go down the hall, I floated down the hall. When I tried to stop, the part of me floating just stopped. I remember looking at the family room couch, the television, and the bar my dad built. I was struck by how *normal* everything looked.

I paused in front of the back door, wondering if I could go outside. In a blink, I was outdoors, viewing the world from above the back porch.

The moon was the only bright spot in the dark sky. The wind ripped past me, but I didn't feel it. I knew a fall chill was in the air, but it didn't hit my skin. The eaves were at eye level. They were packed full of leaves in various stages of decay.

Of all the things I could see in this state—of all the places I could go—my only thought was to visit my girlfriend. I didn't stop to consider the why or how or the absurdity of it all. I just went.

Sue lived in another part of London, Ontario, in a two-story side split. I followed the laneways and roads to her parents' house as easily as if I was driving my motorcycle. It seemed like the natural thing to do. I remember floating fifteen or twenty feet off the ground and gliding over cars and trucks. Houses flew by. The buildings looked stubby and unadorned from this strange perspective. I'd never realized air conditioners were on rooftops. I had no clue how fast I was traveling, but I passed moving vehicles.

Everything looked different from above, yet I wasn't afraid. My body burned with excitement, and my mind was on overdrive. The feeling was exhilarating.

The downstairs lights were still on when I arrived at my girlfriend's house and there were cars in the driveway—Sue had mentioned her family was hosting friends that night. I wondered how I would see her, then decided to go around the back of the house to the bedroom she shared with her sister, Vicky.

The bedroom lights were also on, and I could see Sue through the window. She was sitting on her bed beside her sister. They were talking and laughing, and I wanted to know what the joke was. Without thinking, I moved my head through the wall below the window to get closer.

Sue looked my way and screamed. Her sister followed suit. They both gawked at me before jumping from the bed and running from the room, terrified.

A wave of fear washed over me. *What have I done?*

I floated home in a panic. I think I followed the same route, but can't be sure. This time, I wasn't interested in the view. I couldn't believe I'd done such an awful thing to Sue and her sister. I entered the house, relieved, releasing my breath at the back door.

That's all I remember. How and when I got back to my room, my body, is a mystery to me.

When I woke the next day, I remembered every detail of the night before. I had no explanation or even a theory, and by later that day, I'd convinced myself it was a dream—impossible but all-too-real. It had to be, or shame would eat me alive. I'd never intentionally frighten Sue and her sister.

That night, I headed over to Sue's house after work. We had plans to see a movie. I arrived on my motorcycle, and her mother let me in the front door. "The girls are in their bedroom," she said, grinning.

"What's so funny?" I asked.

She couldn't contain a laugh. "Those two up there think they saw a ghost last night."

I must have taken the stairs three at a time, but stopped at the door to my girlfriend's room, suddenly worried about the reception I was about to get. My heart was pounding. I knocked a warning on the door and opened it to step inside.

Sue and her sister jumped from the bed.

"You!" Sue yelled.

I just stared. What could I say?

"Something came into my room last night," Sue said, pointing toward the wall beside the window. She shook her head in disbelief. "I swear it looked like you." Her sister clasped her hands over her mouth.

I was about to hurl. I didn't know what to say or do, and the girls were just as frazzled. My girlfriend gaped at me like I had fifty heads. She wasn't far off. I wasn't normal. I saw strange things, spent entire evenings with alien beings, and could travel outside my body. All I had to do was say so—to throw my hands up and confess.

I took a deep breath and laughed out loud.

"We're serious," said Vicky. "The ghost looked like you."

Had I not been nineteen, immature, and scared out of my mind, I might have reacted differently. If it was 2021 and not 1976, or if I wasn't an expert at keeping my secrets hidden, I might have confessed all and made an ally or two.

Instead, I laughed my fool head off, making light of the enormous rock that churned in my gut. I told the girls they were crazy. I suggested they were delusional, seeing things. Denial was my best course of action. It was my only course of action.

Within minutes, I was out of there, fabricating a reason to cancel our date. Telling all would lose me a girlfriend. The truth would make me the laughingstock of our social group. I met my buddy at a tavern that night and drank away the shame.

I never had that dream again—the one where I was a boy falling down the black hole. I believe I learned the lesson it was meant to teach me.

We are not our bodies. We are our consciousness.

6

THE UNFATHOMABLE

It was as if the Ewok-like creatures had largely disappeared in the fall of 1976. Perhaps they'd moved on to another child. Maybe they got what they needed or wanted from me and I was of no further use to them. I have no idea. If they were part of a bigger picture, a team or group of extraterrestrial creatures working as cohorts, their job was done. I don't have the answers. What I do know is that after seeing the vision of the man surrounded by light and experiencing the night outside of my body, I'd changed. I felt relief and a new enthusiasm for life. But it didn't last.

There was no warning. I hadn't felt an alien presence in my room. No vibration coursed through my body, and there was no blue-green light taking me to a room of white. I woke on a cold, hard examination table. I was pinned to it. I couldn't feel or see restraints of any kind but I couldn't move my hands or feet, or lift my head. My mind was spinning. I could not see my abductors, yet I immediately knew this was a new alien experience.

I'd never been captive like this before. I had no idea if I was clothed or stripped. I could hear, but the sound of my heart beating in my ears muted any outside noise. The only smell was the scent of my fear. This place—whatever and wherever it was—didn't look like anything I'd seen before.

Panic hit me hard and fast. And the pain was unbearable. I hurt *every-where*. I couldn't see what was being done to me, but I could *feel* it. The lower part of my abdomen was on fire. I was in and out of consciousness.

Something moved from the left into my line of vision. For a brief moment, it loomed over me, tall, hairless, and inhumanly thin. Then it disappeared. I stopped breathing. *What was that thing?* It was nothing like the Ewok creatures I was accustomed to. A movement to my right caught my attention—another similar creature. It was hard to focus. My head was thick with fog. I felt drugged.

From my horizontal perspective, one alien was a foot or two higher than the table I was on, and the other towered over me by at least four feet. I didn't see tools or equipment, but they were working over my torso. Both beings were extremely skinny, and the tall one had long, thin arms and what resembled three long fingers at the end of each arm. Both had long, tapered necks topped with huge heads. Their heads were shaped like teardrops—about a foot wide at the top and four or five inches wide at the bottom. There was a small hole on each side of the head, which I assumed were ears of some sort. Their mouths were slits about two inches wide. They didn't have lips, and their mouths didn't appear to move. Three inches above the mouth were two small holes about a quarter of an inch across and maybe one inch apart. If this was a nose, it was flat to the face and didn't move to pass air.

The tall one's eyes were large, tapered like a kernel of corn, about four inches apart and positioned a few inches above the holes I thought were the nose. They looked too big for its head. Although the creatures didn't look me in the face, I could see their eyes were jet-black and shiny. They were cold, black holes showing no depth or emotion.

Unlike the Ewok creatures, these beings didn't communicate with me or each other in any evident way. They didn't care that I'd woken up on the operating table. They were oblivious to my screams. My feelings were none of their concern.

My entire body radiated with pain. I was convinced I would die on the table. I tried to see—I remember registering bits of their heads, faces, and bodies as they came into focus—but I couldn't focus. Blackness came and went. I have no idea how much time passed.

It would take several of these abductions for me to be able to describe these creatures adequately. Between the pain, possible sedation, and unadulterated fear, I had a hard time making sense of the experiences. They registered in my mind like flashes of a horror movie I could never forget. I put them together like puzzle pieces and as the years passed, I could better describe what was happening to me.

Their skin was a deep gray, of varying shades. Some were lighter, some darker. This skin wasn't smooth. There were wrinkles like small frown lines that ran horizontally on their bodies. They didn't wear clothing, coverings, or jewelry of any kind. I couldn't see any hair. And I never saw their feet or legs, so I don't know what they looked like from the waist down.

Over time, I realized these creatures were accompanied by shorter creatures who had a different air about them—they looked more human. Their mouths were slits with small lips, and the center of their face had a bump that could have been a nose. Their eyes were also dark and unmoving but more rounded. I'm not sure what their purpose was or why they hovered so close during these procedures. The tall ones did all the work.

These abductions continued for well over two decades—about once every three to six months, until I was in my mid-forties. It might have happened more. I have no way of knowing. I only remember the times I woke gasping in pain on an operating table. It's possible that I was sometimes taken without being conscious of it.

To these creatures, I was merely an animal to be probed and tested. I have no reason to believe they were interested in me personally or trying to help me in some way. It was as if my sole purpose was to satisfy their curiosities. When I woke during a procedure, it was never for long. Something put me back to sleep. I prefer to think it was their doing and not my body reacting to the unfathomable. I like to think they cared enough to try to keep me unconscious.

I was subjected to indecencies and violations I will never understand. My mind checked out; the trauma was too much to bear. I struggle immensely with these bits of memories, much more than any other extraterrestrial experience I've had. So much so there were moments I felt I couldn't write about these years, these experiences. While I have no injuries or physical scars that would hamper my quality of life and I'm grateful they didn't kill me, which they easily could have done, the dive into these encounters is extremely painful.

By all accounts, I should be deranged. At the very least, I should be unstable. But I'm not. Sure, I don't like to sleep without nightlights, and I have a hard time sleeping alone. For years, I've piled pillows around my body when sleeping—something I can't believe I am confessing. I don't know why, but the pillows give me a sense of control and security, even though I'm well aware they do nothing to stop abductions.

Despite all this, I've refused to let these events define me, make me, or break me. I've forced myself to find the positive within the negative. I clung to humor, empathy, and kindness as ways to counter the helplessness I felt during abductions. Not only did these encounters make me a mentally stronger man, but they also helped me to appreciate the good in my life. And there was a lot of good.

Along the way, I completed my on-the-job training and passed the tests to be a certified electrician. After working and learning in the field for several years, I wrote and passed the Red Seal Examination to be a licensed electrician. I would later become a licensed Master Electrician and spend many years working in an industry I adored.

Life wasn't always easy, especially when my father was diagnosed with terminal lung cancer. I had to pause my education while that was going on. Dad was having anxiety attacks that scared my mother and sister. He insisted the cobalt therapy made him suicidal. He passed away from cancer in less than a year. My mother would follow less than ten years later, also dying from cancer. These trials gave me motivation, a vested interest in making my life mean something. I had big aspirations.

Sue and I didn't last. I dated other girls over these years until the summer of 1986 when I went water-tubing with my roommate June and a bunch of our friends. June brought Lisa along, convinced we would hit it off. Lisa worked part-time at the dental clinic where June was a dental assistant. At nineteen, Lisa was almost ten years younger than me, yet I was smitten from the first moment and I managed to catch her attention that first time we met. She had dark hair, an athletic body, and hazel eyes that stopped me in my tracks. There was something about her, how she was comfortable in her skin and knew who she was, even at nineteen. She laughed at all my stupid jokes and thought I was kind. June's matchmaking skills were impeccable.

Like me, Lisa was hardworking and goal-oriented. I loved that she was a spitfire with plans to be a lawyer. Despite her traditional WASP parents thinking I was a motorcycling hooligan, too old for their daughter, we stayed together, determined to have the careers and family we wanted.

In April 1991, five years after meeting, Lisa and I had our first child. I was thrilled to be the father of the precious baby boy we named Joshua. Lisa and I were married in 1992, and two years later, in April 1994, we welcomed our daughter, Keira, into the family. I was built to be a dad. Doting and overprotective, I spoiled the kids at length. I still do. I didn't have a lot growing up, so I want my kids to have everything. I got to be the good cop while Lisa set the rules.

By day, I was a father, a provider, and a husband. Life went on. I was an adult. I had responsibilities. I needed normalcy like a person needs air.

But by night, I fought the demons of past and present abductions. While I wanted these encounters to end, how could I fight them? What choice did I have? I couldn't tell anyone about the abductions. No one would believe me. I couldn't tell Lisa; she and the kids were my escape, my reprieve. There are some things you can't share with your spouse. Besides, what could she or anyone do to stop these beings from taking me? I also feared that sharing my experiences with my family might somehow implicate them in what was happening to me. The last thing I wanted was for anyone else to get abducted.

The more precious my world became, the more I feared that sharing my secrets would crush everything I'd worked so hard to create. At best, I'd be a joke to my coworkers and neighbors. More likely, I'd be shunned by family and friends and stripped of my electrical career. I feared being locked up, probed, and tested by my own kind. In no scenario did I come out a survivor. I maintained my silence.

It would take over twenty years and the coming of the Internet for me to realize I wasn't alone, that others had been taken from the comforts of their beds, rooms, offices, and cars in cities and countries across the planet. I lived decades thinking I was alone, a one-off, a freak of nature. Today, these innocent victims refer to themselves as "abductees" or "experiencers." They have lived through things only a few can relate to, and even fewer can speak of.

They are not crazy. They are unprotected prey. Most refer to their captors as "The Grays."

7

UNKNOWN LANGUAGE

In 2007, I was fifty years old, living with my family in our dream home in Newcastle, Ontario, just outside of Toronto. My wife and I had just celebrated our fifteenth anniversary. I was a master electrician and worked at one of Ontario's largest nuclear power plants. Lisa was a successful lawyer and partner at a civil litigation firm in Toronto. She specialized in insurance law and was still the motivated self-starter I fell in love with. Josh was sixteen and a senior in high school. He couldn't wait to be an engineer or biochemist and had the marks to get there. Keira was thirteen and well on her way to being a beautiful young woman with a strong head on her shoulders—like her mama. I was still the one the kids turned to when they wanted something, especially a free lunch. Lisa made sure the kids did their homework. I was the softy.

On the night of July 28, 2007, Lisa was away on a work retreat for the weekend. I had the bed to myself. At around three in the morning, a loud thump woke me from a deep sleep. It was coming from outside, along with a bright light that shone through the French doors of our bedroom. I got up and looked out over the cedar fence in our backyard toward the muddy field of grass and weeds that separated our neighborhood from Highway 401. The light was gone. I could see nothing but pitch-dark night. I crawled back into bed.

I'm not sure if I fell asleep or not, but I was wide awake when my body began to vibrate to the point of frightening me. It had been a long time since I'd felt this kind of vibration, this pull. It was more like the Ewoks than the Grays, and I'd forgotten how it felt. I could see and hear, but I couldn't control my body's movements. I was paralyzed.

From my position on the bed, I could see the outline of a being standing beside my nightstand. I couldn't make out a face in the dark. It wasn't more than four feet tall, but nothing about it reminded me of the creatures from my youth.

In a deep, scratchy voice, it started to talk to me in a language I had never heard before. Unfamiliar words ran together like a stream of consciousness with no beginning, middle, or end. It made no sense to me, but I could hear it speak directly inside my head, much like I did when the Ewok-like beings spoke.

Confused and shuddering from the vibration, I managed to say, "I don't understand." The creature paused a few seconds, then started talking again. It wasn't clear whether it was talking to me or another being. I couldn't see anything else in the room.

If this creature did anything to me or took me anywhere, I don't remember. The only thing I recall was listening to the creature talk to me in that unknown language.

When the vibrations finally stopped, I could move. A light was again shining in from the back of our house. The being was no longer in my room. I could move. I jumped out of bed to peer outside.

Two creatures were walking toward a light. From what I could see, they seemed human, or human-like, only very small and slender. They walked and moved in a familiar manner. They didn't look furry, but there was hair on their heads. They were wearing black or dark outfits of some sort.

I have no idea what possessed me to follow them, but I rushed down the stairs and flung open the patio door to stand on the deck. The two entities approached the light. I could now see that it was coming from a craft hovering about five or six feet from the ground. There were no moving parts, no wings, no windows, and nothing to show this thing could move. Still, I assumed this was their mode of transportation. It was the first time I saw a spaceship from this perspective. My usual view involved a blue-green light coming up from below.

I could see the craft was oval, about twenty feet wide and ten feet high. Something opened on the ship—a door, I assume—and my backyard lit up like the sun. I had to shield my eyes from the stunning light, which was making it difficult to see the creatures disappear inside.

Moments later, the ship rose into the air without a sound. It paused about a hundred feet above the ground before disappearing in the blink of an eye.

These were creatures I'd never seen before and have never seen since. I didn't know why they were there or what they wanted. All I knew was that if there was an intergalactic list of earthlings that extraterrestrials deem fair game, I was on it.

8

REMOTE VIEWING

This encounter with this new species of extraterrestrials piqued my curiosity. Childhood memories came flooding back. I tried to piece together all the abductions I had endured, all the procedures I'd been through. I wanted to make sense of them. I decided to focus on my near-death experiences to see if they might unlock some clues to the other phenomenon.

By mid-September 2007, I was chin-deep in near-death stories I'd found on the Internet. It seemed I wasn't the only one to step outside my physical form on occasion. Thousands of people reported these out-of-body experiences. Many of them were ridiculed relentlessly.

While searching the Internet, I came across some information on "remote viewing." This was a concept I'd never heard of before. While remote viewing wasn't easy to understand, I got the basics. It was defined as the human ability to acquire accurate information about a distant place, person, or event, without the use of physical senses or obvious means. Even the US government put weight on this concept. I read article after article about military protocols related to the US government's interest in remote viewing. The more I read, the more I learned. Experts claimed remote viewing was associated with clairvoyance, and some referred to remote viewing as an intuitive second sight or sixth sense. In rough terms, it was knowing something without any logical reason for knowing it. Apparently, it was useful in espionage.

I was hooked.

I wondered if remote viewing was an explanation for the near-death experience I'd had when I was seven and the out-of-body experience I'd had at nineteen. Or if there was at least a connection. I didn't have prior knowledge of the concept, and I wasn't sure I believed it was even possible for a person to envision and understand things they couldn't see in front of them. The skeptic in me was unconvinced. But I was curious. What if I had somehow tapped into this sixth sense?

I found a website that offered a CD on remote viewing. I managed to download it and began practicing the protocols. The key, it seemed, was learning to clear your mind of any thought. This sounds easy enough but it proved to be hard to do. My mind was always racing.

Still, I committed to giving remote viewing a shot. I was excited to see where my mind could take me.

Ignoring Lisa's glares and my kid's quizzical glances, I loaded the CD and chose my first blind target on the screen. Being a blind target meant I couldn't see the image, only a number representing the image and an otherwise blank screen. The instructions said to use a blank piece of paper and a pencil, which I did. I scribbled my name on the top right corner of the paper—a step meant to initiate focus. I cleared my mind as best I could. I tried not to think of anything.

Once my head felt clear, I concentrated on whatever thing would be next to enter my consciousness. As a vision crossed my inner eye, I outlined what I saw, drawing it on the blank sheet of paper.

At first, I thought I was seeing the blade of an airplane. But as the image became clearer, it had a distinct tower and four rotating blades. My art skills were lacking, but the end result was obvious. I'd drawn a windmill in a field.

With bated breath, I clicked the button to reveal the blind target, the picture I hadn't seen before this moment. It was a picture of a windmill with a distinct tower and four rotating blades in the middle of a field.

My mind was blown.

I stayed with the protocols laid out in the CD for about a month. I mentioned my fascination to Lisa and the kids. They thought it was nuts but interesting. They didn't understand the context as I did and had no idea why I'd be intrigued by such a thing.

Sometimes the blind target tests worked; sometimes they didn't. I'd estimate it was a fifty-fifty split. Then, one Friday night after work, I had the house to myself for the evening and decided to change things up. Instead of using the CD, I tore four random pages out of a magazine without looking at them. I tucked each page into a separate large envelope, then wrote a number on the top right corner of each envelope, along with my name.

I had no idea what each envelope contained, but I picked one and set it in front of me. I was sitting at my desk and had the lights dimmed to help me relax. The house was quiet. I cleared my mind of all thoughts.

What happened next was remarkable. It was like I had a big-screen television in my head. I saw a Harley Davidson motorcycle in the center of the screen and writing all around the bike. It was clear as day, as if I was holding the magazine page in front of my face. My eyes were closed, but I could look around the advertisement and even read the words.

After two or three minutes, I opened my eyes and tore open the envelope. The magazine clipping was an ad for Harley Davidson motorcycles. The colors were vivid, the image was precise, and the wording was exactly as I'd read with my eyes closed.

I slumped in my chair, stunned. I couldn't believe it.

Again, I cleared my mind, took another envelope from the stack, and proceeded to repeat the steps. When I closed my eyes, once again, a large screen appeared in front of me. At first, it was empty, white, a blank canvas, but soon something was being filled. The image came to me like an Etch-A-Sketch on fast-forward. It was a shoe, a woman's high heel shoe.

Again, I opened the envelope and was astonished to see an ad for women's footwear. Front and center was a black-and-white drawing of a lady's high heel shoe.

How could this be? How was this possible? Did I have a super-power?

Weeks later, I tried again. I followed the same instructions and went through the same steps. This time, the tests were unsuccessful. I couldn't clear my mind. Maybe the gift had come and gone. Either way, I didn't give up. Every so often, I pulled out the CD and gave it another go. Sometimes it would work, and sometimes it didn't. Whatever talent I had was unreliable. I wasn't going to be recruited by the government to uncover secret documents any time soon.

What I learned is that the human mind is capable of much we don't understand. This opens up so many possibilities. What if we are capable of bilocation: the ability to be in two places simultaneously? What if consciousness, defined as the quality or state of being aware of some-thing within oneself, is much, much more than those simple terms? What if we are more than genetic strains and outward appearances? What if we're more than biological beings? What if we contain wonders just waiting to be discovered?

9

SAY CHEESE

By January 2008, the US housing collapse had toppled markets through-out the world, including Canada, sending people into a panic. I was sent to work at the Bruce Nuclear Generating Station on Lake Huron's east-ern shore in Tiverton, Ontario. There'd been planned power outages that needed attention, so Bruce County, lovingly known by Ontarians as the gateway to cottage country, was flooded with power plant work-ers, including electricians like me. By the end of training day, I was on a wild search for accommodations that didn't exist. The area hotels and inns were packed, and most locals willing to rent a room had already found a tenant.

While driving from hotel to hotel, house to house, and adding my name to already long waiting lists, I stumbled upon the Skylite Motel in the small town of Kincardine. With all honesty, the place was decrepit. The shag carpeting had stains I refused to think about, and the old orange curtains were pocked with holes I hoped were sunburns and not bug bites. The room was so small that the single bed swallowed the entire room. The bed, too, was small. My feet would hang over the end. But that was the only room left, and as crappy as it was, it was better than spending days or weeks sleeping in my truck. I took it.

I was in better spirits than I should have been, considering the room. I was stoked to try my newly purchased Moultrie D55ir trail camera.

I'd attempted to catch my extraterrestrial visitors on video before—a few years before—but failed. The Sony camcorder I bought could tape while I slept, but the timing was limited. At best, I could get an hour and a half of continuous recording. This meant I had to wake every hour or so to rewind the tape and recharge the battery, which became tiresome. The nights I could even try the camcorder were limited. I couldn't set it up at home, where I shared a bed with Lisa. We'd been married sixteen years, but I still couldn't bring myself to tell her my most guarded secret. Lisa had no idea I'd been abducted by aliens the vast majority of my life, even during our time together. I knew she was my escape, my refuge, my tether to a life of normalcy. For my own sake, I didn't want her to know. Another part of me refused to put such a burden on the woman I loved. Lisa was a savvy, analytical lawyer, and I had no proof, nothing concrete to show her. When I played back the camcorder recordings, there was nothing but darkness; all the lights would need to be on to see anything. And I couldn't sleep in those conditions—not for days or weeks on end, waiting for my abductors. I'd be too tired to function. So I'd given up trying. There was no way to capture these creatures on camera.

Then, just before Christmas, I'd overheard a bunch of work guys talking about hunting deer. They were planning a hunting trip and were organizing supplies. I'm not a hunter. The idea of killing anything for any reason doesn't sit well with me. I don't have it in me. But I was intrigued when I heard the guys planned to use something called a trail or game camera to scope the local deer population. While the technology seemed unfair to the deer, I was fascinated by the guys' accounts of night pictures being somewhat reliable and clear. Trail cameras are motion-activated, taking photographs as the deer or wild animals pass the camera. They use an LED flash and infrared to capture subjects at night.

Eureka! I'd found a way to take pictures while asleep.

This crappy little room at the Skylite Motel was my first opportunity to test my newly purchased trail camera. I was jumpy with a combination of excitement and anxiety. I pulled the camera out of the box and set it on the table beside the bed. It was an ugly thing. The outer casing was hard and painted to camouflage with the bark of a tree. There were four rows of LED lights along the top and a camera lens in the middle. It wasn't light—it felt heavy in my hands—and was way too large to carry around. After reading the instructions, I inserted the batteries in the back and programmed the SD card, preparing the camera to take shots.

I got ready for bed in a flash, despite the slow-running water, lack of heat, and limited lighting. But when I positioned the camera and hopped into bed, my thoughts were laced with fear. *What if I'm caught?* This was a hotel room, not a forest. The camera stuck out like a beacon, screaming, "I'm up to no good." What would happen if aliens understood I was catching them on film? Would they allow me to take pictures? Would they get angry? Would I be risking my life?

I hardly slept that first night. I tossed and turned, and the camera flashed and clicked when I moved in my sleep. The next day, when I downloaded the pics from the SD card onto my laptop, there was nothing but a man trying desperately to sleep.

Maybe this wasn't such a great idea after all, I thought.

I knew they were coming—the extraterrestrials. I just had this sense. Several months had passed since my last encounter, and my internal alarm was ringing. I had to be patient. If a lifetime of extraterrestrial experiences had taught me anything, it was that these creatures would come. They could find me anywhere.

The first night was uneventful. After the second, I woke remembering a little of the evening's events. All I recalled were the vibrations and seeing the blue-green light. But I knew in my gut that something had happened, that I'd been visited. I jumped out of bed, anxious to see if the camera had caught anything.

The trail camera had taken several pictures in the night, and the black and white photos were fairly clear. I scanned the pics, hoping to find something, anything, that could collaborate my experiences. I felt deflated, seeing only a man in bed asleep.

Then, in one of the pictures, I saw it. The perspective was skewed, unclear, yet something tangible.

The creature had been moving past my bed—or possibly I'd moved in my sleep—when the flash was triggered, and the camera caught the top of the alien's head as it approached the lower end of the bed.

I couldn't believe it.

The Ewok-like creatures had returned! They weren't done with me after all. These beings had visited me so many times in my youth; I would recognize them anywhere. There was no doubt in my mind I'd found proof, actual evidence that my abductors existed.

I wasn't sure what to do with this newfound information, and it ate at me for days. Not only had I caught a being from another planet on camera, but I also lived to speak about it.

That weekend, I returned to Newcastle, glad to be home. I wasn't myself. I was exhausted, distant, and agitated. The kids were in their rooms and Lisa and I had just got comfortable on the sofa, about to watch a movie when something in me exploded. I hadn't thought it through or made a plan; it just blurted from my mouth like an active volcano. "I have something to tell you about my childhood," I said.

Lisa didn't even look my way. "Spill it," she said, flipping through the television channels.

My skin bristled with live nerve endings. My stomach hurt. I started to tell her about childhood abductions that began when I was seven. I described the Ewok-like beings and how they took me from my bed—whatever bed I slept in. Slowly, I explained how these creatures seemed to teach me things, although I struggled to recall the details of what they taught me.

"Are you sure?" Lisa asked when I paused for air. Her hazel eyes were wide, sincere. She'd shut off the television and dropped the remote at the first mention of the word alien.

"Absolutely." I am many things, but I am not a liar. I would never make something like this up. Lisa understood this. After twenty-one years together, sixteen married, she knew me. "These weren't nightmares or daydreams," I added. "These creatures were real, living, and they took me whenever they pleased."

I couldn't bring myself to tell her about the Grays. I just couldn't. She was already looking at me in awe, overwhelmed. She was shocked. I watched her gather her game face, the one she uses when hit with an obstacle at work.

"Did you tell your parents? Did they know?"

I thought about that for a moment. "There were occasions I think they saw or heard things related to my experiences, but they didn't understand. I told them many times in the beginning when the abductions first started. But they didn't believe me."

Lisa believed me. A massive weight was lifted from my shoulders. I finally had an ally, a confidant. I had to control my emotions for fear that tears would spill from my eyes.

"So, this is why you've been so distant lately, depressed. Have these . . . things hurt you?" Lisa didn't seem frightened—a condition I had no desire to change. If I didn't show fear, maybe she wouldn't either.

I shook my head, *no*. That wasn't a lie. I had never been permanently harmed by aliens, at least not in any way I could prove. Mental scars were something altogether different, but I wasn't about to mention those.

"What are we going to do about it?" Lisa said.

What could we do? Nothing. And now I was torn. I had finally shared my deepest, darkest secret with my partner, the woman I had vowed to share eternity with. But in doing so, I'd burdened her with a weight no person should have to bear. I'd gained a confidant but lost ignorance, innocence, and my reprieve from the madness.

The king of denial was now in purgatory.

We agreed to keep my secret between us. There was nothing to be gained by telling the kids. We didn't want them to be afraid in any way.

I told Lisa I'd bought a trail camera to take pictures in the night. She stared at me, uneasy. In hindsight, I should've kept quiet about the trail camera, knowing Lisa would never warm to the idea, but I wasn't thinking straight. I showed her the picture I'd been carrying in my pocket like it was a piece of government identification, proof of my existence in a world where aliens come to visit. She couldn't see what I saw. How could she? She could see the strange thing beside the bed, what might be the head of something only a few feet tall. But, having never seen an extraterrestrial, she had nothing to compare or relate to.

Lisa didn't want to talk about it anymore. She believed me, but my experiences raised questions about the nature of reality that she found disturbing. My secret was too much to digest.

Honestly, I couldn't blame her. I knew the mantra all too well. *Deny, deny, deny* . . . I'd spent over five decades pretending I was normal, that creatures from other worlds didn't exist or abduct me. Now, my wife would pretend right along with me.

This is the trail camera, the Moultrie D5Sir I purchased to catch night photos of extraterrestrial visitors.

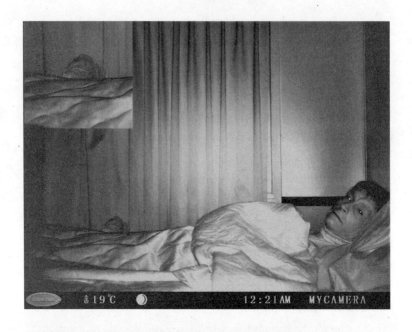

10

STRANGE THINGS

Some days, I leave for work at two o'clock in the morning. When my job takes me to Tiverton, Ontario, I'm in for a three- to four-hour drive—in good weather.

One Monday in winter 2008, I headed out amid a snowstorm. In twenty minutes, I managed only a few city blocks. I decided to turn around and head home. It was safer to wait for daylight.

I took a second shot at 9:00 a.m. and, this time, I managed to get about an hour outside of Toronto before the visibility got so bad, I had to exit the highway and pull into a rest stop. I was ordering a coffee when I overheard a police officer warning the staff that the county was closing the roads. I had two choices, either sleep in my car or find a hotel and hunker down. I chose the latter.

I settled for a room at an inn in the next town over and ordered room service. The walls were thin in this old building. I could hear the couple talking in the room next to mine. They sounded young. Their fast-paced chatter about the storm was interrupted by the sound of opening and closing drawers. I suspected they were stranded like me—only they had luggage.

I checked in with my boss and called Lisa, then spent the day watching television shows I'd never seen before. I ordered food. Around eleven o'clock, I turned in for the night. The bed was comfortable, and I curled onto my side to watch the storm outside the window. I had no trouble falling asleep.

At some point in the night, I was rocked awake by an intense vibration. On instinct I tried to flee, but I had no control over my muscles. My mind was racing. My body wouldn't follow commands.

The couple from the room next door started talking again. Only this time, it sounded like they were in my room, standing behind me, beside my bed. I was sure of it. I desperately wanted to turn my head but couldn't, and my limited sight range was blinded by darkness.

"I am going to set you up."

I had no idea what this meant. It took me a moment to realize this voice was feminine and sounded nothing like the woman from the other room. This voice was deeper and slower. I didn't have time to panic. Something moved on the bed, and I was grabbed from behind. My legs were pushed over the side of the bed as I was lifted into a sitting position. I couldn't speak, but the vibrations told me all I needed to know. I wasn't being moved by a human being.

"Are we coming as well?" said the woman from next door. She didn't sound afraid. If anything, she sounded excited.

The creature didn't answer her—not in a way I could hear. "Do you want to fly?" she whispered in my ear.

Did I want to fly? Fly where?

My eyes—the only body part I could move—searched the room. There was no blue-green light and nowhere to go. There were no holes in the ceiling or a spacecraft hovering. I could see the door. A small amount of light filtered in from the hallway.

The creature pressed against my back as she wrapped her arms around me from behind. I think I gasped. I tried to scream. Her long arms felt bony and thin. She squeezed so hard I had trouble breathing. "Are you ready to fly?" she asked.

"Are you kidding me?" I thought. "No!" I could do nothing but stare into the abyss.

She rocked us off the bed, toward the floor, and I think I yelled out in shock, expecting to go up or out—not down.

It happened so fast. The floor rushed toward my face. My sight went pitch black. For seconds, I think I felt a falling sensation, but I can't be sure. I remember nothing past this point.

I woke the next morning in the same hotel room. The bedsheets were on the floor. It hurt to breathe. My sides were tender and visibly bruised.

I thought about knocking on the wall to check on the couple next door but couldn't bring myself to do it.

Deny, deny, deny.

I had experienced a fair number of crazy things in my life. But this encounter was strange.

11

SEARCH PARTY

In October 2008, I was working on a transformer station in Ottawa. The days were messy and long, and my crew and I spent the nights camped at a local Comfort Inn near the site. We'd only been on this job for a short time, but a typical routine had been established. When we headed back to the hotel at the end of the workday, we looked forward to a hot shower before dinner and a cold beer.

On this particular day, I stumbled into my hotel room around 6:30 p.m. The minute I stepped into the room, something felt off. I tried to ignore the slight vibration running through my body. It made no sense. I was accustomed to the vibrations at night, but the sun was beaming through the window. I was wide awake.

I sat on the edge of the bed, peeling myself out of the dirty overalls I had worn all day. The vibration seemed to pause or disappear and for a moment, I wondered if I'd imagined it was there at all.

I shimmied my overalls to my thighs before bending to unlace my boots, and wham, the vibration hit me like a blunt object to the head. I saw stars.

That's all I saw. I remember nothing more.

When I didn't show up for a drink or dinner, the guys were worried I hadn't made it back to the hotel. They were sure I'd said I would meet them at the restaurant attached to the inn, where we always got together.

They went to look for my car in the parking lot, hoping I hadn't had car trouble or an accident.

My car was there, parked where I'd left it. Fearing something worse had happened, they banged on my hotel room door. They knew I'd been struggling with health issues related to a weight problem that started in my late forties. I was scheduled for gastric bypass surgery in New York later that month.

When I didn't answer the door, the guys, perplexed, went around the building to peer into the window. My room was on the first floor, and the cleaning staff always opened the curtains when they made the beds. The guys could see my entire room. The lights were off, and my bathroom door was wide open. There was one of my work boots on the floor—which they thought was strange—but I was nowhere in sight.

The search party called it a night. They had no reason to believe I was hurt in any way.

I woke in my room at 6:30 the next morning. Exactly twelve hours had passed, yet I was lying, face-up, on the bed, still wearing my dirty work clothes. My feet were hanging over the edge of the bed, one boot still laced. The other boot was on the floor, right where the guys had seen it the night before.

Confused and dizzy, I noticed the time and scrambled to make it to the morning brief, where I was expected by 6:45 a.m. I grabbed some clean work clothes and raced to work.

The guys were relieved to see me. They told me all about their worries the night before. They recited all they'd done to find me. They'd even left messages on my phone throughout the evening, concerned about my safety.

What could I say? My palms were sweaty. I felt ill and off-balance. I didn't know where I'd spent those twelve hours, and my suspicions were unspeakable.

I lied through my teeth. What choice did I have?

The guys had a hard time believing I'd called a cab and went to another restaurant alone—unshowered and still in my work clothes. The story was lame and didn't ring true. I'd never done such a thing, and what about the boot? But it was the only lie I could think of on the spot and under pressure. I was beyond exhausted, and I was pretty sure I hadn't slept a wink.

The guys reluctantly let it go. I could see they were curious but uncomfortable. If I hadn't looked so bad, they might have pushed for more, but they didn't. Unshaven and sullen, I was a pathetic sight.

Whatever happened in those twelve missing hours, I doubt it was good. I'd never been taken for an entire night before. At least, not that I could recall. It was frightening to think of aliens probing or experimenting on my body for hours and hours. I had no physical injuries I could see or feel, but I was still rattled.

I tried to put it behind me, to forget. I didn't want to remember such things.

Self-preservation is a powerful instinct. The mind finds ways to survive, to exist without collapsing under a weight too heavy to support. For weeks I walked on eggshells around the guys. I wasn't myself. When they asked questions, I clammed up. Eventually, the suspicious glances subsided, and I was grateful when things got back to normal.

Or as normal as my life could ever be. How many people lose entire blocks of time?

Apparently, the number is higher than I ever imagined.

Dwarf Doctor at my bed side in Newcastle, Ontario.

12

GOOD HANDS

Not all my extraterrestrial encounters were scary. In fact, I believe some creatures wanted to help me. Unlike the Grays, there were alien beings who showed an interest in my feelings during abductions. They had an ability to control my level of anxiety without use of drugs or sedation. It seemed they wanted me to feel as comfortable as possible or, at least, less afraid.

This made sense when I thought about it from their perspective. The less I struggled, the easier their mission. Whenever I struggled against their mind control and reacted negatively, they knocked me out, or my body shut down of its own accord. Either way, as the years passed and I survived encounter after encounter, I suspected there was a method to their madness, even if the plan was beyond my understanding.

My family and I had plans to celebrate my fifty-second birthday in a few weeks. The milestone held meaning since my gastric bypass surgery. My weight had dropped drastically. Dangerously, in fact. Within twelve weeks I was down to less than half my total body weight. As a result of this sudden loss, my body attacked the muscles in my heart, something the doctors referred to as broken heart syndrome. March 29, 2009, I was rushed to hospital with heart failure.

I was lucky there was no long-term damage, but the experience frightened me. Afterward, I had to be on an IV drip called Total Parenteral

Nutrition (TPN) to supply my body with nutrients it couldn't produce on its own. I would later discover the botched gastric bypass caused more than heart failure. Nothing makes you take note of your delicate mortality like major surgeries.

This particular spring day was like any other, although warm. I remember getting up at six a.m. and turning on the air conditioner. Lisa was sleeping. She would be up soon. She worked nine to five and took the train to her law firm in the city. I was dressed and out of the house in under twenty minutes.

I drove to my usual coffee shop for my morning coffee and cream cheese bagel. I noticed the trees with their various shades of spring green seemed to glow against the cloudless blue sky. I got to work with time to spare, read the local paper, and proceeded with what turned out to be a good day. I got a lot done. I remember feeling quite pleased with myself, considering I was still recovering from surgery. My lethargy seemed to be lifting. That evening, my wife and I watched television before I headed to bed early. That's when my nice, normal day ended.

That night, I was taken to what appeared to be a land base of some sort. I woke on a moving gurney that was moving up a ramp into a wide hallway. The large ventilation pipes coming out of the floor caught my attention. I'd never seen piping or ventilation aboard an alien craft. The floor looked like poured concrete. The doors had hinges and doorknobs, which also struck me as strange. It was almost like being in a regular hospital.

I noticed I was being wheeled up the ramp by what looked like an elf. The being had pointy ears and a round, human-looking face. I could see its shoulders and head from my horizontal position. The overhead lights were dimmed, but if I could've lowered my hand to the floor, I would have touched it. The gurney was only twenty or thirty inches from the ground. This creature was short.

I was wheeled into a massive room where another elf approached, its ears also quite long and pointed. This one had kind eyes—human eyes, only different somehow. His nose was larger than any I'd seen before, but there was no doubt it was a nose. I remember thinking the little guy was handsome. He was wearing a white lab coat that hung past his knees.

"Hi," I said, unsure what else to say. "Nice to meet you."

He looked startled at first, but a smile crossed his face.

"Likewise," he said.

That was the extent of our conversation. Immediately, I fell asleep.

A few weeks later, I had again gone to bed early, but was not yet asleep. Lisa was downstairs watching a movie. I'd been tossing and turning, trying to get comfortable. The elf doctor appeared out of nowhere, standing beside my bed. He said nothing, took my hand, and smiled.

We didn't go anywhere or do anything. I didn't pass out or wake in an unfamiliar place. I didn't feel drugged, paralyzed, or controlled in any way. The creature just stood there, holding my hand.

I should have called Lisa, but the thought never entered my mind. In fact, I never had the chance to speak. I smiled back at him, blinked, and he was gone.

I wish I'd had the chance to tell him that shortly after his initial visit, my human doctor discovered that I was healing better. Fast, even. The TPN wasn't needed anymore. My body was absorbing nutrients through my intestines as it should. I was finally on the road to recovery.

Did the elf doctor have anything to do with this improvement? I can't be sure, but it felt like he did. It was as if his visits were for no other reason than to make me feel better.

Maybe I was in good hands after all.

13

MIND GAMES

In spring 2009, I was working at the Darlington Nuclear Generating Station in Bowmanville, on the north shore of Lake Ontario. I was electrical support during routine maintenance on the motor control centers. It was late May. The days were getting longer. The end of a shift didn't feel like the end of the day at 5:30 p.m., especially after I'd showered and shed dirty overalls before leaving the facility. I enjoyed the drive home this time of year. The sun put a bounce in my step and the trip from Darlington to Newcastle was a mere five miles.

I pulled onto the highway. I had a coffee—cream, no sugar. Traffic was moving well for rush hour and I lowered the window to let the cool spring air fill my lungs. A few transport trucks lingered to my right, and a bright yellow sports car passed me on the left. I turned the music up a bit. I was looking forward to an early dinner with Lisa.

I remember looking at the clock on my car dashboard, noticing it was 6:00 p.m. on the nose when I grabbed my coffee to open the lid.

BAM. Suddenly it was pitch dark.

In shock, I slammed on the brakes. The car swerved before I got control and pulled to the side of the road, onto the gravel. Nothing but darkness and the silence of night surrounded my idling car. My headlights were on, which meant my tail lights were on as well. I pried my hands from the steering wheel. My breathing came heavy as I stared at the coffee cup in the console beside me, the lid still unopened. I searched for answers outside the windows, but all I could see was a full moon and the odd set of passing headlights.

The clock on the radio read 9:30 p.m.

Holy shit. Exactly three and a half hours had disappeared in an instant.

I began hyperventilating. My head was spinning with confusion. I hadn't even blinked, and it was suddenly night. How was this possible?

Nausea enveloped me, and I fiddled with the tab on the coffee cup, in need of a calming distraction. My hands were shaking. The coffee was cold. I reached for my cell phone to confirm the time. There had to be a logical explanation. Maybe the clock on my car experienced some strange power glitch? As an electrician, I knew this wasn't possible, but in this moment, I was willing to consider anything but the worst—my worst. I couldn't fathom how the sun could vanish in a heartbeat.

The time on my cell phone read the same. It was now 9:32 p.m.

I could see numerous calls and texts from Lisa spanning the hours. My gut reaction was to call her back, but I hesitated when I realized I had no idea where I was. What could I say? Lisa was already worried about me being back to work. I'd recently had a heart attack. The gastric bypass had reduced my weight from almost 400 pounds to about 150 pounds in eight months, putting enormous stress on my body and heart. I couldn't worry her more. Instead, I prepared to drive and eased onto the highway in search of something to signal where I was.

The first sign my headlights hit said "*Welcome to Brockville.*" I started to shake again. I was familiar with Brockville. I'd lived there when I was young. Brockville is further down the highway. Much further, about three hours or 175 miles further.

I dialed home.

"Where the hell are you?" said Lisa, answering on the first ring.

The words weren't coming easy. "I'm in . . . Brockville. I have no idea how I got here. I can't explain it."

Lisa paused. I was sure she was calculating the distance. "Are you all right?"

"I was driving and the day just . . . disappeared."

"How did you . . .? Why would you end up . . .?" Lisa went quiet. She believed in aliens to the extent she couldn't imagine we were the only lifeforms to exist in the universe. She believed that I was truthful about the experiences I felt comfortable sharing. But she'd never seen an extraterrestrial with her own eyes, and she didn't worry when I didn't worry. *Deny, deny, deny.*

"Are you okay?" she asked again.

"I think so." A quick body scan didn't reveal anything notable. "I don't feel sick or hurt. I'm baffled, but I feel okay." I felt jittery, nervous, but I wasn't hurting anywhere.

"Find a place to stop. I'll come to get you." Lisa knew my parents' reactions to my childhood abductions had a huge effect on me. She was determined to be stoic and supportive from a protective distance.

My head was trying to process what had happened, but all I could focus on at this point was getting home. I pulled off the next exit. "I'm turning around," I said. "I'll be home as soon as I can."

I only got a few exits before realizing I was almost out of fuel and had to pull off to find a gas station. My head was churning with possibilities. Could I have had a seizure? Is it possible for someone to have a three-and-a-half-hour seizure and stay driving without hitting anyone or crashing? If I was abducted, how could the car keep driving, moving miles down the highway, without me at the wheel? Did the car really continue down the road, or was it placed here, with me in it? Do aliens take entire vehicles? Could I be driving on the highway and be somewhere else at the same time? Could I exist in more than one place or dimension? Why would I not remember one or the other?

I had the long drive home to fret over the ordeal, replaying events in my mind. Not once did I come up with an explanation that made sense. Of course, my mind jumped to alien involvement or unusual phenomenon—all options were on the table, given my past. But even that didn't offer a reprieve. This was nothing like the time before when I'd gone missing in the hotel room and my work buddies searched for me.

The more I thought of it, the more my stomach roiled. I recalled a time when I was twenty-nine and driving Lisa's 550 Suzuki motorcycle from London to Muskoka to meet her stepfather at his brother's cottage. Lisa and I hadn't been dating long, and I was eager to make an impression. Her family needed help with work on the cottage, and I'd offered my electrical skills.

It was early summer, mid-June, I think. I'd been enjoying the ride—Lisa's bike was fun to drive. The trip to Muskoka was a good 200 miles. I'd already clocked over three hours due to stop-and-go traffic through Toronto when I pulled off in Washago for gas. With only twenty minutes to go, I was anxious to get my ass off the bike and get my hands dirty.

I'd only been back on the highway for a few minutes when I blinked and was suddenly in bumper-to-bumper traffic. I didn't have to slam on the brakes; I was just there, in a different place, as if plopped out of time. I recognized the area almost immediately. I was back in Toronto, almost two hours from Washago. I remember thinking, *what the hell?*

People didn't drive around with cell phones in 1986. While I could have left the highway to find a pay phone, who would I call? I didn't have contact information for Lisa's stepfather, and Lisa was at work. What could she do for me anyway? What could I possibly say? The truth would make me look crazy.

When I finally arrived in Muskoka, Lisa's stepfather was frantic. He thought I'd broken down on the highway and was heading out to find me. I brushed it off, suggesting he was confused about the time I was scheduled to arrive. I don't think he believed me for a minute, but he wasn't about to challenge me. I was there to help, and there was work to do.

When I got into the cottage, I searched for a clock. Sure enough, hours of my day were missing. *Up in smoke.* It was an anomaly I couldn't explain at the time, and I'd never experienced anything like it before, so I let it go. How could I ever explain something so strange? I couldn't get the pieces to fit in my own head.

All these years later, I still struggle to understand what happened in those missing hours.

I've since learned that a substantial number of abductions take place in moving vehicles. There are hundreds, if not thousands, of reports with similar details. People worldwide have reported being taken from their cars or trucks, either stationary or in transit. And while most abductees recall their experiences in vivid detail, I've also learned that many report hours or days of memory loss—time having passed without any recollection. Some recall the abduction experience later when something triggers the memory. A few remember through hypnosis or therapy.

I've also learned that memory or time loss isn't only experienced by the abductee. While the media often suggests the majority of abduction experiencers are taken when alone, this is not true. Most abductees are taken with witnesses. Studies suggest that almost 80 percent of abductees were not alone when taken, and up to 62 percent of witnesses can recall part or all of the event in detail. Husbands and wives are often present while their spouse is experiencing an encounter or abduction in the night. Yet, for some reason, they don't wake during the traumatic events of their loved ones. Or, if they do, they can't recall it happening.

How and why does this happen? Why do some abductees and witnesses remember events and others don't?

I think alien beings have the ability to control our memory to a certain extent. Maybe this is why abduction memories are so often revealed under mild hypnosis. It's possible an outside source merely masks these memories. Remembering strong feelings such as terror, rage, sadness, and pain can be debilitating. And while I can't be sure, I don't believe alien beings want us fearful or anxious. If they did, why go to such great lengths to help us forget moments of severe trauma? By hiding or masking traumatic events from our consciousness, our fear is removed.

Now when I think back on my alien encounters, I can't say if my lack of fear came from my perspective changing over time, from learning aliens don't hurt me in any permanent way, or if I've been externally stripped of my worries. How could I know? It is possible aliens have somehow altered how I think or recall these events, removing fear from the equation with some sort of anesthetizing energy from their minds, hands, or rod-like instruments.

I believe, over time, those like me who remember traumatic experiences and snippets of time, somehow break through the memory control. The memories I do keep are quite vivid and real.

Maybe it's best we don't remember details. Perhaps we aren't capable of coping with events so incomprehensible our minds can't process what our bodies have endured. Had the experiences that happened to me in my late teens and twenties been the first or only experiences to take place, I doubt I would have been able to handle such atrocities. The Grays were not kind. Many abductees claim their childhood encounters changed for the worse in their late teens when their bodies hit puberty. Then, the abductions became rougher and more physical. At the very least, I'm sure I would have reacted very differently. Since I'd been abducted many times as a kid and was more or less unharmed for several years, my coping mechanisms were fine-tuned by the time I fell into the hands of the Grays.

Perhaps I should be grateful.

If I was abducted by the Grays on that summer day when driving Lisa's motorcycle to Muskoka, the missing time or memory is probably a good thing, something I don't want back. The same goes for the lost three-and-a-half hours on the 401 in 2009.

Sometimes, for me, ignorance is bliss.

14

BILOCATION

I was at home, in Newcastle, and I'd just gone to bed. I wasn't tired but the news had me stressed. Haiti had just been hit with a massive earthquake, and hundreds of thousands were missing or dead. President Obama was sending help from the United States. Our Canadian Prime Minister, Stephen Harper, was following suit. I listened to music with my headphones while waiting for Lisa to finish putting dishes away downstairs.

Both French doors were propped open, allowing the hallway light to cast a faint glow across our otherwise dark bedroom. I couldn't see much in or around the room, yet I had no reason to think or feel aliens were present. My internal clock was ticking without a care, and there were no vibrations or butterflies in my stomach.

Something started scanning my body. It felt like a slow, narrow line of air blowing on me with medium force. It started at my toes, slowly making its way up my body as if I were a piece of paper on a photocopier bed. It didn't hurt, and I wasn't overly afraid, but I'd never felt anything like this before. I freaked out and jumped out of bed.

Well, I tried to jump out of bed, to escape whatever this scanning sensation was, but I couldn't move. Again, I was paralyzed.

The scanning feeling passed over my face. I don't recall if I cried out for Lisa or if I was too confused to attempt to yell for help. The feeling reached the top of my head, and shocked me with a loud snap. I was instantly in the arms of what appeared to be a nine- or ten-foot-tall being.

I was looking up at his large face. He looked masculine and human, but for his height and strength. His head was much bigger than mine. He was looking straight ahead and not at me. His arms were massive, and I could see and feel them around me, effortlessly carrying me through some sort of wheat field. The sun was shining in a clear blue sky. There were evergreens and massive trees around the field's outer edges. It looked and felt like earth, like home. I thought we were in a farmer's field. I remember being confused and panicked.

In another instant, my eyes were open and I was back in bed. Music was filtering through my headphones and my room looked as it always did. But something didn't feel right. I still felt as though I was in the arms of this giant man. My arms and legs were right where I could see them, but they felt like they were bouncing with the rhythm of the man's footsteps. The sun's heat was still touching my face. The breeze was moving my hair. I hadn't gone anywhere, but I had.

I was aware of both realities. I was in two places at once.

I closed my eyes tight. Again, I could see this giant man and the field of wheat. My body moved as I was carried through this field, the man not showing any sign of connection.

I'm not sure how many times I opened and closed my eyes, but the sensation abruptly ended. My music was still playing, and I was in my bed.

Lisa entered the room and got ready for bed. I sat in silence, unsure of what to say. Even when she lifted the sheets to join me, I couldn't utter a word. I was disoriented.

What the hell had just happened?

Within minutes, my wife was sound asleep. I stayed awake most of the night, trying to make sense of things.

I now believe *they*—some sort of extraterrestrial race—were showing me that it's possible to be in two places at once. We are capable of bilocation. The giant man, however, still confounds me. If he was human, he was the largest man in existence. If he was not human, he certainly looked the part. But I don't know if we were even on earth.

15

TECHNOLOGICAL DIFFERENCES

The moon is covered in crater holes, extreme weather scars, and volcanic fissures that prove it has been inundated with catastrophes. Mars and several other planets mapped by mankind show similar histories. Earth is no different. Our planet has been hit with constant change since its creation.

It's not hard to imagine how ancient civilizations could have experienced major extinction-like events that wiped out all or most traces of their existence. There is strong scientific evidence that humans thrived and crashed along with all other lifeforms on our planet at various points in earth's history. It's a big deal to think of our ancient ancestors were mapping astronomy and developing medicine and science instead of living primitive existences. Some of them were. We know this. We've come a long way as a species. Probably more than once.

The chances we are the only species to do so are slim. The idea we're the only intelligent lifeform in a universe filled with billions of planets is ludicrous. Every species, however, would be in a different stage of development. Time and galactic setbacks could, in theory, create a vast array of living things with unique abilities.

Evolution is a bumpy ride.

I've seen a fair share of technology as an electrician for Canada's power plants. I understand how machinery works and how modern equipment is powered. I also know there are clear differences between human technology and the machines used by those who abduct me. While I usually struggle with some combination of mental fog and stress that makes it difficult to concentrate on my surroundings during abductions, the details that have struck me over and over, becoming almost familiar, have always stayed with me.

For example, it's hard to forget round or curved rooms that appear to have no depth, no visible doors for entering or exiting, and light that is ever present but seems to have no source. One remembers equipment and instruments that seem to grow from the wall or floor as if formed organically or melted together. My eye is drawn to the missing corners, seams, and joints. Out of habit, I look for wires and cords leading to power sources that are never there. Their equipment doesn't have dials or buttons or handles that require manual dexterity. Screens and markings don't exist. Machines make few sounds I recognize as electrical. Materials seldom resemble plastic or wood. Surfaces, in general, are usually shiny, smooth, and cold to the touch.

I assume their spaceships are nothing more than modes of transportation, but I could be wrong. They could serve other purposes, have meaning beyond points A and B. Either way, they don't look like our cars, boats, trains, or planes. Most of the ships I've seen are no larger than a transport truck. They don't move like our vehicles or show obvious signs of propulsion.

My surroundings during abductions are clearly not made by human hands. In fact, I imagine our five human senses—sight, sound, smell, taste, and touch—have limited relevance to extraterrestrial technology. Light, in our world, does not cut through physical matter, erasing everything within it. It's not capable of lifting weight or manipulating gravity. Walls don't disappear as we walk through them into bare rooms where equipment is suddenly conjured, as if by magic. Our machines don't work without touching them or commanding them. Even with my vast experience as an electrician, there's been little I could relate to, and even less I understand about alien technology.

From my perspective, select creatures from afar are further along in their evolutionary progression, and their understanding of technology is leaps and bounds different from ours.

The same could be true for all living things in our universe.

On November 14 and 15, 2010, I was staying in a rental house in Port Dover. It was a nice but small place on the water, and two work buddies and I had been crashing there for weeks while working at the Nanticoke Power Plant. There were two bedrooms of reasonable size upstairs and a tiny bedroom on the main floor. I took the small one.

The first night I was abducted, I woke in midair. My hands were still tucked behind my head and my feet were crossed. It felt like I still had the security of my bed beneath me as my body rose with the blue-green light. The vibrations were strong but tolerable. My pajamas were gone, as usual. I could see the bed linens in a pile on the end of the bed far below me but couldn't move much. I was physically and emotionally controlled. I couldn't see my abductor through the light, but I could feel his presence. He was close, watching me.

As I rose toward the ship's bottom, I could see it was maybe 100 feet in diameter, silver, and round. I couldn't tell how high above me it was hovering—my perspective from below was skewed, and the blue-green light saturated everything it touched. The silence was deafening. If the night was breezy, I couldn't feel it on my skin. I didn't even feel the November chill. The houses below me were getting smaller by the second, so I must have been moving pretty fast. I felt calm, slightly sedated, and the panic that threatened to engulf me when I saw the neighborhood and lake become dark blotches in the night came and went pretty quickly.

My memory after this point is blank. If I made it to the ship conscious, I don't remember entering any door or hatch. If I was probed, tested, or taught, I have no recollection. I woke the next morning as if nothing had happened. Every inch of me felt fine, normal, and I went about my day.

The second night was nothing like the first.

An intense vibration woke me, racking my body from top to bottom. Fumbling upright in bed, I could hardly see my surroundings. The usual blue-green light was intense. I could scarcely make out the clock on the night table. It was 3:00 a.m.

Of course, I thought to myself; *they are right on time.*

I was surprised they'd come two nights in a row. That didn't happen often. I also found it strange I could move. My body wasn't paralyzed or pinned, and although the vibration made it difficult to maneuver on the bed, I was able to kick off the sheets and shimmy to the edge. My nightclothes were still on. Usually, by now, creatures had stripped me bare. I searched the darker recesses of the room for my abductor, my tour guide, the one responsible for taking me up the light and into the ship. No one was there.

A shiver ran through me. I seldom traveled the light alone.

I was wide awake and clear-headed. When I looked up, I could see the light had cut a round hole through the ceiling. The upstairs bed-rooms, my roommates, the furniture, and the roof were nonexistent in this space. The light flickered, and I noticed it was awkward and differ-ent. The green tinge was missing. I ran my hands through the bright blue light while a sinking feeling came over me and hearing an unusual hum, like the fading note of a plucked guitar string.

I sensed an alien presence I didn't recognize.

I felt a tug. My pajamas lifted first, taking my arms and legs with them. It was as if my wrists and ankles were locked in cuffs I couldn't see. They hurt as I was lifted from the bed. The rest of me hung unsupported. It took all my effort to keep my head up.

Higher and higher I went. All my weight gathered in my back, putting immense pressure on my lower spine. My aching joints were stretched to their limit. I remember it was hard to breathe in this folded position. I think I cried out—I must have. No one answered, either verbally or mentally. This felt wrong, like the entire system could shut down any minute, dropping me. It felt crude.

The beam of light wasn't thick like usual. I could see the clouds and an ominous moon.

I couldn't support my head any longer. It fell back, and above me the spaceship came into view. Small, less than fifty feet wide, maybe half the size of the ship from the night before, it was rectangular, perhaps square. I couldn't tell if it was silver or white. A strange wave of light moved from one side to the other, in time with the humming. It reminded me of those desk toys that rock back and forth with slow-moving liquid in a plastic box. The surface was smooth and reflective, except for small bumps that covered the surface where the beam met the ship in a circle.

Something hit me, maybe an invisible vibration or radiation source. I was suddenly struggling for air. My heart was pounding. I felt doped to the point of nausea, like I'd been shot with some drug. This was not usual. I usually passed out before getting this close. Then everything went black.

I woke on a cold table. The room was small, maybe twenty by twenty, and also bright. Usually, the walls were rounded or unseen and lit from within, but here the light came from a strip down the center of the ceiling.

I felt heavily sedated, and my limbs and head were difficult to move. I could hear my breath coming in heaves like I was struggling to breathe, yet I couldn't feel my chest or lungs. A vague recollection of someone having grabbed my arms and legs and lifted me onto a table or gurney came to mind, then the thought slipped away, unattainable. My eyes couldn't focus. The table felt hard, metallic. My body was shaking, but not from a vibration. I think I was in shock. The table was too small. I felt like I was going to fall. I tried to stabilize myself, but couldn't get a grip.

As I struggled to keep from falling and to stay conscious, I heard movement. Nothing shifted within my limited vision. To the left, I could make out a wall of cabinets. I'd never seen cabinetry on a spacecraft. They were darker than the room, metallic, maybe silver in color. There were matching taps below the set of cupboards—a sink? It looked like a sparse old doctor's office. There were no pictures on the walls, no equipment I could see.

Something entered my field of vision, then disappeared before my mind could register details. I heard running water—something I'd never heard during an abduction—and worked to turn my head toward the taps and sink. My eyes fought to stay in focus. Someone was moving in front of the cabinetry. This creature was tall, around six feet. I could make out the back of a long jacket of some sort. The creature's shape was familiar, its movements reminding me of home, and when I was able to focus on its head, I could see the back of a full head of striking blond hair.

This creature made me think: human.

16

NOVEMBER 15, 2010

OTHER HUMANS

I managed to grip the sides of the cold table—trying to keep from crashing to the floor. I was surprised to see a full head of blond hair on an alien. I'd never seen an alien with human-looking hair. Other than the Ewok-like creatures, which are covered with fur like an animal, I'd never encountered an extraterrestrial that wasn't bald. I found myself opening and closing my eyes, squinting, trying to correct my vision, but the scene before me didn't change. And when he turned around to face me, stepping close, I gasped.

This creature looked like a man. From this close, I could see his hair, although blond, was no shade I'd seen at home. It was brighter, a shocking yellow hue. He was nordic-looking, and rather young. His skin was smooth and healthy. He had two eyes, a nose, and a mouth, positioned like ours. I don't remember noticing his eye color. His build under the coat was that of a fit thirty-something man. He looked . . . perfect.

Almost too perfect, yet polite and nonthreatening. I wasn't nervous in his presence. He leaned over me, cocking his head to get a closer look. I think he was trying to understand what state I was in and whether I was conscious. Heavy sedation made my limbs hard to maneuver, and I couldn't keep my eyes open for much more than a few seconds at a time.

He said something to me, but I couldn't make sense of it through the brain fog. Whether he spoke using vocal cords and his mouth, I couldn't tell. Despite the sedative, I knew this was the first time I'd ever been abducted by a creature who looked human. Part of me was relieved this creature was man in some way. Part of me was petrified to consider what that meant.

The man turned back to the cabinetry, and it dawned on me my limbs were not restrained in any way. Could I escape? Where would I go? I was still fighting to stay conscious. I could hardly move my head, never mind run.

There was an open archway on the right side of the room. I'd never seen so much as a door during an abduction, so this archway got my attention. I could see it led to a larger room with gray floors that resembled cement. Like the room I was in, the walls were white and lit from above. Two teenage boys were sitting on the floor facing each other, cross-legged. They were dressed in what looked like human clothing, nothing that drew my attention as extraordinary. They both had pale skin and a shock of blond hair. I was too far away to note if they were twins, but they certainly looked similar from where I lay. Perhaps they were the doctor's kids or relations. They reminded me of my son Josh.

Movement caught my eye. I watched as one of the boys rose from the ground, legs still crossed, until he was levitating about three or four feet in the air. He was jittery, his body shifting as if unstable.

I'd never seen an alien float. He remained in position with his hands on his knees as I tried to stay conscious to watch. He appeared to be staring at the boy in front of him, the one sitting on the floor in the same stance. I could hear them talking. Each had individual voices not yet broken by manhood. They sounded like any other teenager I'd heard on earth. My kids, Josh and Keira, were sixteen and nineteen at the time.

The one floating said what sounded to me like, "There's a sale on windows at Home Depot."

Home Depot? For obvious reasons, this struck me as a crazy thing to hear on a spaceship. Yet, I'm absolutely sure they were talking about home renovations and Home Depot. They spoke for several minutes, casually, as if their surroundings were normal and home renovations were standard fare. This seemed absurd to me, even in my state.

The man-alien running water caught my attention when he cleared his throat—a very human thing to do. It took effort to turn my head in his direction.

"Is there anything you want to ask me?" he said.

This time, I clearly made out every word he said. He spoke with his mouth, using English. He crossed his arms and smiled. There were no tools or instruments in his hands, but I knew what was about to happen. Dread clawed at my stomach. I wasn't in any condition to be the one making demands. I couldn't think straight.

"Will I feel pain?" I finally mumbled.

He didn't shift his body weight or touch me in any way. He didn't move closer or turn to grab something from behind. The room went dark anyway. I did not feel any pain.

Were these creatures human? Like, really human—our DNA and all? I have no reason to think this. Your average human being doesn't shine a blue-green light on your bed, cutting through tangible matter to lift your body to a spaceship in the sky. Humans don't generally have bright blond hair and a perfect complexion. We also don't levitate. Yet, I suppose our race could have evolved at one or more points in time. Who am I to say there weren't ancient civilizations of man, people who invented ways to escape the catastrophes that annihilated our planet thousands or millions of years ago? Maybe our ancestors took an unrecognizable path to survive. If we look beyond what we think we know today, and what we choose to believe, anything is possible.

Maybe these human-looking lifeforms are a more advanced version of us. They could be us, genetically, only different. It's even possible they aren't more advanced in all ways, only some ways. This would endorse the theory that humans are the product of accelerated evolution. This popular concept suggests humankind is so much more advanced than other life on earth because we've been tampered with biologically. Mankind has taken some pretty big leaps within the last few centuries. Some think this is not a coincidence—we had help.

Of course, theories are just that. I can only hypothesize.

While I didn't see or sense aliens during this abduction, there is a chance these humans were abductees themselves. The one in the long coat appeared in control, but his life prior to this day could have taken any number of turns. What if he was a human raised by an alien race? Or a helper, a man born of earth then thrust into alien employment or partnership? Many abductees talk of seeing other human beings during abduction experiences. Some recall seeing human children or adults within a spacecraft or otherworldly room at the same time they are taken. Many speak of human assistants among the spaceship's crew. Maybe this doctor and these teenage boys were humans who lived or worked with aliens?

I've never been taken alongside another person, but that isn't to suggest it can't or doesn't happen. Abductees are taken in pairs and groups. Sometimes married couples are abducted together, like Barney and Betty Hill in 1961.

I've also never seen a young child of any race on board an alien ship, not even when I was abducted as a kid. Yet, research suggests this is not the norm. Child abductees are often encouraged to play with alien kids and other human children or hybrids on the spacecraft during abduction events. It's not clear if this is to make the abductee comfortable, assist in the development of the alien or hybrid child, or allow aliens to study human children for research purposes. But it's certainly a common thread among abductees.

My perspective only covers my personal experiences. When considering the possibilities, my mind must stretch. Could the human-looking doctor be an alien/human offspring, a genetic modification? I'd be blind to disregard the large number of experiencers who claim they've been part of genetic engineering experiments to create hybrid offspring—beings containing both human and alien DNA.

For most of my adolescence, I was visited and abducted by the Ewok-like creatures, almost always involving the female who referred to herself as my mother—the nurturer, the one who kept me calm. I always assumed she referred to herself as my mother because she compared herself to my biological mother or thought the presence of a mother figure would comfort me. These were assumptions, however. And maybe I was totally off base.

What if she was my biological mother? Would a hybrid know they are a mixed breed? Apart from my rare and unique blood type, I'm a pretty normal guy, and this isn't something one tests for. Earth's population could be dotted with genetic mutations without us knowing. If aliens have mastered gene manipulation and mixing to some extent, I'd assume they'd be pretty damn good at hiding it. This concept scares the hell out of me.

Why would aliens create hybrids? I don't know. Modern movies would have us believe aliens use us for sinister purposes, like replenishing their genetic stock or creating controllable armies within our ranks. But this theory would contradict the vast majority of my extraterrestrial encounters, which were quite benign. My guess is that the treatments I was undergoing with the Ewok-like creatures were to adjust what we call our DNA. John E. Mack says it best in his book, *Abducted*. "My impression is that we are witnessing something far more complex, namely an awkward joining of two species, engineered by intelligence we are unable to fathom, for a purpose that serves both parties."

If creating a hybrid race is indeed their goal, there must be a really good reason to go to such lengths. What if it's to save us? What if we are in galactic danger of becoming extinct, and aliens want to preserve our kind? What if a drastic change in our DNA is the only way we'll survive into the future? The reality is, we do this every day when we take steps to stop earth's wildlife from extinction. Who's to say we aren't another race's pet project?

Nearly all abductees recount events that included intrusive medical procedures. They are stripped naked and taken to labs. Like me, many are paralyzed or sedated during these experiments or procedures, and some regain consciousness while aliens are doing things to their bodies. Aliens use instruments to penetrate their bodies through the nose, eyes, ears, vagina, anus, and more.

Abductees worldwide have been returned by their captors with odd burns, rashes, cuts, marks, and lesions on their bodies. Long, thin bruises are often found on women's calves, thighs, and buttocks, suggesting finger marks. Patterned bruises or abrasions are common on the arms and chest of abductees. Sunburn-like rashes and burn marks that cover the back or entire body have been documented. Restraint marks are found on wrists and ankles, and puncture wounds are common on or around the hands, feet, and genital areas. These physical signs of abduction are often small and undiagnosed by human doctors. Most disappear rather quickly, and some leave scars that modern medicine can't heal or explain. Even abductees with no memory of the experiment or surgery itself struggle with knowing something has been done to them—something beyond their control and without their consent.

I can't stress this enough: Even abductions that don't result in permanent physical harm are extremely disturbing events. I consider myself lucky. My extraterrestrial experiences have not resulted in any sort of physical scarring. For the most part, I've been taken and returned unscathed. I have no doubt my body has been used in ways I prefer not to think about, but the physical proof has been limited. Perhaps this is because the abductions started when I was a child and aliens came to know me—we bonded. It's possible I've never met the criteria for further study. Or, maybe I didn't get hurt because I don't fight back. I can't say why my experiences are different than others. Either way, I'm relieved I haven't had to explain physical ailments or scars to my wife or family. Most of my scars are internal.

But not all. When I was young, in my teens and early twenties, I used to find weird things on my body. To be fair, I reckon most kids at this age find their bodies change in ways they don't understand. A lot happens to the human body when maturing. Hormones are out of whack. Body parts morph into things we don't recognize, seemingly overnight. It wasn't until I was much older, a full-blown adult with teenagers of my own, that I realized I might have mistaken abduction corollaries for biological maturing.

For years, I used to get these dots on my chest and arms. There were always three red dots about three millimeters each and perfectly round. They sat exactly an inch apart in the shape of a triangle. There was never more than one of these triangle spots at a time. I used to find them on the left side of my chest, closer to the front, or on my left arm, between my elbow and shoulder. There was never any bruising or swelling around them. They were flush to my skin and didn't hurt. They would remain for a few days, then fade away. I'd look a few days later and they'd be gone.

The same pattern appeared dozens of times over the course of five or six years. Some I noticed after an abduction and some didn't seem to coincide. I would catch them in the shower or when I was getting dressed. I made every effort to pretend they didn't exist. Although I didn't know what caused the dots, I was always worried someone would see them. They made me self-conscious enough to hide them from my parents and siblings. I never told anyone.

The red dots weren't the only abnormality. A few weeks before my seventeenth birthday, I woke with patches of hair on my forearms. They were thick clumps about two inches wide in the shape of a chicken egg. When I rubbed a hand over an arm, I could feel the drastic texture change between my wrist and elbow. The hair was thicker than my usual hair, and my normal blond hair was missing from these patches. But what disturbed me the most was their lack of color. The strands were transparent. They looked more like fishing line than body hair and didn't resemble anything else on my body, new or not.

The patches appeared on the same spot on each arm. I hadn't noticed them before showering that morning, and a comprehensive body search didn't uncover patches anywhere else. I covered them and went to school but couldn't stop touching them when no one was looking. I was kind of freaked out, worried someone would see them. How could I keep my arms hidden from my girlfriend? I knew my sisters would tease me. I remember thinking how seventeen was no age to have strange body hair. I had to get rid of these hair patches.

My bright idea? I got tweezers from my mother's bathroom and pulled the hair out. They were hard to remove. Yanking them hurt like hell—tears came to my eyes—and some even bled.

Once my arms healed, the hair patches never returned. I considered the experience bizarre but explainable. My father's friend had given me an old snowmobile to fix up, and I'd removed the hood to rewire and clean the guts. I'd something to my arms while fixing this snowmobile, I figured. I must have got chemicals on my skin, causing strange hair to grow. It could happen, right?

I was losing a lot of hair back then. I had a motorcycle and would find hair in my helmet regularly. I was worried enough to take the advice of a girl whose friend got help at a local hair clinic. They took a scalp sample from my head and recommended a special shampoo. Sounded simple enough. I wouldn't dare tell them about my arm hair fiasco.

Only recently have I learned that a large percentage of abductees report a drastic change in their extraterrestrial experiences around puberty. Like me, many begin their abduction journey in early childhood, and the experiences become less about verbal and visual education and more about physical exploration as they enter their teen years. While most abductees speak of procedures meant to alter or contribute to their spiritual energy or vibration, some abductees feel they've undergone reproductive procedures.

I was shocked to discover it's common for the Grays to get involved at this stage. A large percentage of experiencers speak of reproductive testing or experimentation where sperm is forcibly taken from men and ova are removed from women. It's not clear how these are used or mixed with alien DNA. Still, abductees claim hybrid embryos are reinserted in women's uteruses during subsequent abductions, where they gestate until removed. Hybrid fetuses then grow to full term in tanks or cylinders. Women have been asked to hold these developed babies and nurture them while aliens watch. Sounds very sci-fi. Creepy.

I do not have a single memory of being used or probed for fertility or reproduction. While this doesn't mean these types of procedures didn't happen, I can only say I don't remember any.

That said, I am not naïve enough to think aliens put me to sleep or altered my memory without reason. They did something to me all those nights, and I imagine these things were invasive. Not recalling such things is probably in my best interest. I sympathize with abductees who have endured aliens probing their bodies in the best of times, for reasons they deem tolerable, never mind scary surgeries and sexual tampering at the hands of non-human creatures. My heart aches for these people.

17

TRANSLUCENCE

After that initial week in 2008 when I stayed at the Skylite Motel in Kincardine and tried the trail camera, I often returned to work at the Bruce Power Plant. I rented an attic room in the home of an older couple. I'd work Monday to Thursday and go home to my family in Newcastle for three-day long weekends. The older couple and I became close over the years—they always made me feel less homesick. Their attic was my home away from home.

The room was small but clean. There was a bed beside the door and a window overlooking the backyard. The window air conditioner worked well in the summer months, but the thing was loud. It took a while to get used to. Since I was there so often, I kept supplies in my room: a bag of clothes, toiletries, and my tablet and books—stuff to keep me clean and entertained.

It was there, one night in late spring 2011, when I was startled by a vibration throughout my body. I was wide awake when it happened, propped against the headboard in bed, watching a television show on my new tablet. The light above the bed was on, and I was already in my sleeping t-shirt. It couldn't have been later than ten in the evening.

A strange energy pulsed through my body. This was nothing like the vibrations I'd felt over the years when aliens were close. This was a pulsating energy that hurt. The closest comparison was the pins and needles feeling you get when your body's blood supply to the nerves is cut off from sitting or sleeping awkwardly on a limb and you move or apply weight. All my muscles contracted, cramping.

I'd been experiencing a slight version of this sensation for months, at random times, but this was the first time it didn't fade away after a minute or two. This time, the pulsating got stronger and was accompanied by a strong static-like sucking sensation, as if a mega vacuum hovered above me. The pressure on my head was almost unbearable.

A familiar blue-green light appeared, engulfing me, and I looked down when I felt something stepping on the lower part of my torso.

This creature frightened me. It was no bigger than a lapdog. Its eyes were big and dark like the Grays, only translucent. I could see through its head, torso, and limbs, as if it had no solid mass. The bedsheets and room were clearly visible through its small, skinny body as it climbed over and on me in a rushed, aggressive manner, pulling back the quilt and sheets.

I screamed, yelling at it to go away. I struggled to get out of bed, but the sucking sensation wouldn't let me go.

I felt every inch of my body being pulled as I was lifted out of bed—clothing and all. Above me, the ceiling was gone, replaced with what I assumed was the underside of a spacecraft. The entire bottom was rippling waves of light.

While I remember being lifted out of my bed, I do not recall entering the craft. I have no idea where the translucent creature went or what it did with me, if anything. It was over in minutes.

I woke in my bed with a feeling of dread heavy in my gut. I instinctively understood I was lucky to be alive. It took days for me to shake the experience enough to check my camera.

I'd again been setting up the trail camera before settling in for the night, but now I was unsure I wanted to see what I'd caught on film. When I finally saw the pictures, I was disappointed and exhilarated at the same time. I'd captured the creature on camera, but the blue-green light wasn't visible, and there were no shots showing what happened after I was raised out of bed. The pictures held no answers. They only left me with an endless line of questions to filter through my mind.

How can a being be translucent? How do these extraterrestrials find me? What do they want from me? Why do they come when I'm awake? Wouldn't it be easier to put me to sleep before I know they are present? If they don't care about my well-being, would they care if they killed me? And, ultimately, the question that haunted me the most: what if I'm taken and never returned?

The picture taken this night captured my hair, head, and hands being pulled upward by the light. While the light itself wasn't clearly detected by the camera, the shape of the translucent being on my lap was caught.

18

FAMILY TIES

You would think, considering all I've seen, I could never be shocked.

You'd be wrong. On November 15, 2011, my daughter, Keira, drove me to the emergency room despite only having a learner's permit. She thought I was dying, and I thought she was right. I wasn't in the hospital more than twenty minutes before they had me on an operating table. The gastric bypass I'd had in 2008 had created another major issue. I'd been living with a large section of my intestines twisted 180 degrees and pinned, blocking passage and dying off. Doctors removed the dead part of my intestines and I survived.

Ten days after the surgery, when I finally got home from the hospital, I was in no condition to do much more than sleep. The drugs eased the pain, but they made me jumpy, and I hated taking them. They also kept me awake when what I really needed was rest. This is why I was lying in bed on that particular night, November 25, 2011, with my eyes wide open, when my old friends, the Ewoks, came to visit.

It started with a typical vibration that got stronger by the minute. At first, I was afraid, worried I'd be taken when my body was too delicate to be moved. When the vibration got stronger, however, I noticed that my stitches didn't hurt. I was surprised by this, yet I held my stomach and closed my eyes, just in case.

Lisa was sleeping beside me. By this point, I was used to her being turned off or tuned out so she couldn't see or hear what was happening to me. It was best this way since she'd only worry about my health and safety. The lights were out. The room was dark but for a shimmer of light from the back patio.

I waited, but no blue-green light appeared. Soon, I realized that the usual pit in my gut, the one that told me I was in for a wild night, wasn't present. I could move my limbs freely without restraint. Even my head was clear, despite the human meds.

I don't know how he arrived, but he did. I felt his presence before I saw him shuffle past the walk-in closet to the left of my bed. I immediately knew who this creature was. I felt him. It was the Ewok-like being from my childhood, the one who came night after night until my teen years—the one with the female who called herself my mother. I was overjoyed to see him.

He hadn't changed a bit—although I know nothing about the life cycles of alien beings, so I wouldn't recognize a sign of age if I saw one. He was still small, maybe three or four feet tall, and still covered head to heels in thick, brown fur. As soon as we made eye contact, his vibes of friendship and caring permeated my mind and body. Aliens don't use facial expressions as we do, so I don't know how to explain his reaction to seeing me, but I could tell he was happy.

He approached the bed, his furry head not much higher than the top of my mattress. From here, I could see him clearly. I could tell he was worried about me. *He knows about my surgery*. He reached out to me, his palm open in greeting. I was about to take his hand when I realized there were two other creatures in the room.

I couldn't see them at first, but I knew they were present. I felt them, their slight vibration, their tap on my consciousness—it's the only way I know how to describe it. It was as if they were asking permission to see me, to come up onto the bed. They also radiated kindness, but different than the Ewok beside my bed. It felt like innocence.

I welcomed them onto the bed without uttering a word, yet I was still a little shocked when they climbed onto the right side of the bed, Lisa's side, and bounded over her legs. They were mini Ewoks! They moved like puppies on unsure legs—rough and tumble. I cringed, expecting Lisa to wake, but she didn't. I pushed myself into a sitting position as they nestled beside my right hip. They were very small, mostly fur, but I could see their dark eyes. I couldn't tell if they were boys or girls, but I knew they were kids—his kids. They were miniature versions of their father, the Ewok beside my bed. They sat beside me, one holding my index finger, the other holding my pinkie finger. Their tiny hands were firm but warm.

I was stunned. *Creatures from other planets have children.* Could it be that they have families like ours, including mates and kids?

I felt overwhelmed. I don't know how to compare it to any human feeling. I was inundated with love. They were here to see me, to be with me after my surgery. They knew I'd almost died. They cared enough to come as a family. No words were spoken or thoughts put in my head. We just sat in silence, me in awe.

I'm not sure how long we remained like this, staring at each other, and I don't remember seeing them leave or falling asleep. I woke in the morning sitting up, in the same position, with one hand across my stomach and the other in the spot where the kids had held my fingers. When I closed my eyes, I could see them plain as day—a memory I will never forget.

Like us, aliens have children. Life, nature, finds a way to carry on. Even in the harshest of circumstances, our energy or spirit pulls through. All forms of life on earth, such as mammals, insects, fish, bacteria, and plant life, just to name a few, carry their genes to the next generation in one form or another. Human beings are no exception. Maybe this is true everywhere, even on the far-flung planets of our galaxy. What if all walks of life follow this same generational path? Perhaps this is the one great purpose of all living things.

At the very least, this would explain, in part, the interest aliens have in earth's wildlife, especially humankind. We are, we believe, earth's most evolved being. Our reproductive system is remarkable. Our conscious minds are complex. If I were going to study the lifeforms of another planet, I'd start with the species that won the most chips in the game of life. I would want to understand their family dynamics. I'd be fascinated to learn how similar or different their relationships are to my species. I'd want to know how they work and what makes them tick as a social group. Maybe this is why abductees are often taken with siblings. There are many reports of shared alien phenomena, especially with children, and some alien-human relations continue for generations.

Only a year and a half ago, I discovered my family was no exception. In late 2019, I was at a family gathering. My three sisters were there; it was rare to have us together under one roof. Although we were close in age, each born a year apart, we'd spread out geographically as adults and lived busy lives that made it harder to keep in touch. Since my sister Sharon married my best friend, Jeff, after high school, we spent more time together than we did with Wanda and Linda. This gathering was a chance to catch up, and when Wanda and Sharon were talking, my name came up.

"Robert's writing a book," Sharon said.

"Really?" Wanda was surprised. I'm many things, but a writer isn't one of them. "What would he write a book about?"

"His abduction experiences."

I wasn't beside them, but I imagine Wanda shuffled her feet, surprised. "What abductions? When was he abducted?"

"It started when he was seven. Aliens came into his room."

"Aliens. That's . . . strange," Wanda said. She was startled I was writing a book, yet suggesting aliens had abducted me earned a lackluster response.

"He needs to tell his story," Sharon said, "it's eating him." She thought writing about my experiences was good for me, healthy.

"I understand," Wanda said. What she said next shocked the heck out of Sharon. "I think they came to me first."

Sharon was so surprised by this conversation, she didn't tell me about it until later, days after the family gathering. It was a lot to absorb. Even then, she suggested I speak with Wanda for details.

Wanda believes it started in 1959, when she was four. I was two at the time. She'd sense them in her room. She didn't know what aliens were back then, but she knew they weren't normal, like people, or anything she'd seen before. They frightened her. She'd wake to them touching her feet with cold hands. Then they'd remove her bedsheets. She saw them but couldn't describe them to me. The mind works in surprising ways, protecting us from details too earth-shattering to comprehend. I suspected they were the Ewok-like creatures, but she didn't know or wouldn't say. They would rub her legs, telling her to stay calm. Fear would take over. For some reason, she could move whereas I never could, and she'd run to our parents' room, crying.

She didn't remember how many aliens would come to her room or if they were male or female. She couldn't recall a blue-green light or going to a spaceship. She must have been abducted in that timeframe, yet she doesn't remember, or her memory has been wiped. Or, she remembers and is not willing to confess. While I can't say exactly what happened to my sister during those extraterrestrial visits, it's unrealistic to think aliens would come without a purpose, given my experiences. Wanda said these visits continued regularly for three years, until 1962.

Her answers to my questions were brief and guarded. These experiences were not something she wanted to share. She was extremely uncomfortable talking to me about these memories. She'd shoved them to the recesses of her mind and never thought they'd see the light of day. I can relate. Oh, how I can relate.

Learning this astounded me. It also made me sad. Like me, my sister was a silent victim. For all I know, I was abducted along with Wanda during those visits. I was just a toddler. I remember very little of my life this young. My first memory of an alien encounter was on the night of my seventh birthday. But not remembering previous abductions and them not happening are two very different things.

I couldn't believe it. All this time, fifty-something years, and Wanda hadn't uttered a word of this. Never once did my sister speak up, not even in the beginning, when I told my family what was happening to me. She teased me along with the others. Not once did my parents verbally connect the dots: two kids, years apart, both saying they'd had creatures come to their room at night. Did they think we were just kids having similar nightmares? Did they even ask for details? I can't presume to know what went through their minds, but they didn't say a thing to make me believe they thought our experiences were real.

Since the day Wanda told me her secret, I've tried to talk to her about alien phenomena—hers and mine—several times, but silence ensues. She's not comfortable speaking of such things, and I have to respect that. Her winters are spent in Texas, so we aren't in contact much. Maybe she's mastered denial as I have. Her coping mechanisms are hers to decipher. I don't think she's happy about me writing this book, but she gets my reasons. She's aware her story is included. I think she'd planned to take it to her grave, and I hope I've given her an outlet.

I also talked to my other sisters, Linda and Sharon. Neither recalled being visited or abducted by aliens. While I was relieved to know they hadn't lived through the trials of abduction, I wondered why they'd not been taken, especially considering they shared a room with Wanda. Why Wanda? Why me?

I doubt I'll ever have the answers.

There is only one thing I can think of that makes me and my sisters unique. We all have O-negative blood types. O-negative is quite rare. It's the only blood type with no human history. Scientists have no idea how this blood type came to be, and there are no other species with this blood type on the planet. O-negative blood has no antigens. Our red blood cells have no membrane. This is why we're considered universal donors. O-negative blood can be used in transfusions for any blood type, even newborn babies with undeveloped immune systems. We are, in basic terms, lifesavers.

Wanda and I are anomalies. We are two of only 6.6 percent of the world's population to have O-negative blood flowing through our veins. Could this be a coincidence? Sure. No one knows how or why some human babies are born with this blood type, so every theory should be on the table.

But here's a fascinating fact. Special treatment is required when a woman carries a baby with O-negative blood. Think about this. Without medical intervention, a woman's body rejects a fetus with O-negative blood. Why would nature create a situation where a fetus is regarded as "alien" by the host, the one creating and sustaining the life? The human body considers an O-negative fetus an intruder, a foreign body to be killed off by the immune system. This seems rather contradictory to nature's way of producing and assisting life. Why would this happen?

I don't know. Add it to the list of questions I don't have answers for.

Some people believe O-negative blood may have evolved from an extraterrestrial species. There are doctors and scientists worldwide who believe human beings with O-negative blood are genetically woven with extraterrestrial life. We are, within this presumption, family. While we don't have the science to prove this theory, we also don't have the science to disprove it. Don't believe me? Google it.

I admit the idea I'm genetically related to alien beings is a bit much for me to get my head around. Considering the ramifications is monumental and, let's be honest, beyond my wheelhouse. Yet, I can see the possible connection.

One thing is clear to me. Mankind is not the only intelligent species with family ties.

19

HANDSHAKE HELLO

Back when your computer used a dial-up modem to communicate, the sound it made—a string of beeps, dings, and screeches—was a digital soundtrack that connected servers and computers through public telephone networks. By around 2007, this symphony of sound all but disappeared with the invention of wireless routers that transmit over radio frequencies, bypassing telephone lines altogether. Still, most people today remember a computer's distinct sound when negotiating introductions, otherwise known as a "handshake."

By 2012, I owned the latest technological gadgets like a laptop, smartphone, and tablet. But I hadn't forgotten the nostalgic sound of a computer handshake, and, one day in January of that year, hearing the familiar sound stunned me.

I was stationed in Alberta for an electrical construction project. It was a year-long job on the prairies, and when I wasn't making my rounds as the Canadian Union of Skilled Workers (CUSW) rep, I was crashing at the Canalta Hotel in the town of Oyen. Every two to four weeks, I flew home to spend a few days with my family, then dove back into work. It was a busy time, and personal hours were limited. Most nights I was in my hotel room by seven o'clock, sleeping soundly by ten, and up early to start the process over again.

On this particular night, I was startled awake by an extreme energy surge that pinned me to the bed. I woke on my side and stared at the clock on the bedside table. It was exactly three o'clock in the morning, and I knew there were aliens in my room. I could feel them.

Something moved above me, pushing the pillow against my head and forehead, and I struggled to see but couldn't control my muscles. I panicked. No matter how many times I'd been subjected to creatures from afar, the desire to flee was intense, and there were always moments of terror. Would they hurt me? Would this be the night I wasn't returned?

My head was suddenly pounding to the sound of beeps, dings, and screeches, as if someone, somewhere, was trying to make a connection inside my mind. *What the hell was this? A computer handshake?*

The sound was immensely loud—it pierced through my head. I had no idea where it was coming from or what creature was responsible, but I knew something was moving above me, between the headboard and my head.

I tried desperately to struggle, to reach up, but still couldn't move.

The sound ended for a brief second before starting again, the same pattern of notes hitting cords in my mind. The pressure was unreal, getting steadily stronger, and I screamed, but there was nothing I could do.

This handshake happened three times in a row—each time louder and more painful than the last. Words cannot describe the feeling. Imagine a migraine times ten.

Then it stopped as abruptly as it had started. The pressure in my head was gone. I was no longer pinned to the bed. I leapt from the covers to switch on the light, but there was nothing there. My room was exactly the same. Nothing had changed since I went to bed.

I was so confused. Why would aliens do this to me? What were they trying to communicate? Did they think the human mind was like a computer, and connected as such? Why did the handshake get louder and stronger? Was it because my mind wasn't responding? Was I an experiment that didn't work?

Spending the better part of a year either working or alone in a hotel room results in a lot of thinking time. While alien invasions were the last thing I wanted to dwell on, this handshake had me stumped. I thought about it for months.

While I couldn't say for sure, I didn't think they were trying to hurt me. They were there, above my head, and could have killed me. They didn't. Maybe these particular creatures couldn't communicate as others had, with the use of telepathy or speech. Perhaps they could but didn't know enough about our species to try this route. Maybe I was a rat in a test.

Whatever the reasons, the experience made me think about technology and how—as human beings—we use devices to communicate. If beings from another planet listened to us, if they heard our communication over airways and wires, thinking the human mind is like a computer is not much of a stretch.

Perhaps aliens are more like us than we'd ever imagined. Maybe they are curious, and our kind intrigues them.

Or, maybe—just maybe—they only want to say "Hello."

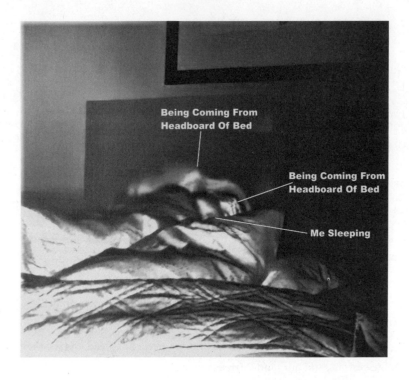

20

STARGAZING

In 2012, we gave up our family home in Newcastle to move into a condo in downtown Toronto. Our lives were changing. Lisa was a managing partner at her law firm—a position she cherished. She was tired of the long, crazy commute to downtown, and rightfully so. She wanted to be closer to the office, cutting her travel time by hours. My work required me to be on the road, so home base could be almost anywhere I could reach by plane, train, or car. At twenty-two, Josh was temporarily back in the coop, but well on his way to independent flight. He'd been focused on an engineering degree with an interest in energy sources and power grids, but had paused his university studies and was working at Lisa's law firm. Keira was almost nineteen and studying at Brock University, where she lived in residence. Her knack for communication was taking her places. Our day-to-day agendas took us in all directions, but we spoke often and gathered for special occasions and holidays.

We didn't stay in the condo for long. The place was designed in levels, and the stairs were hard on the knees. Within a year, we'd moved to a nice house, still close to Lisa's firm.

On a February night, Lisa and I turned in early. She'd been working endless hours at the firm, and I was still stationed in Alberta, flying home when possible. We were both hard workers, devoted to our careers, and a good night's sleep was a luxury. I hadn't been sleeping well for a while. Something had been running around the house in the wee hours of the night, driving me nuts. And we had no pets.

Whatever it was, it wasn't large. While I hadn't seen it, I was pretty sure the thing was no bigger than a bird or small rodent. But it was fast. Its flat or bare feet had been making short but steady slapping sounds on the floor throughout the night, on and off, for weeks. Every time I got out of bed to investigate, the sound disappeared—the creature went quiet.

I'd even resorted to hiding the trail camera in our room—something Lisa wouldn't approve of. She wasn't a fan of the camera. She understood my need to capture this part of my life on film but hated the invasion of our privacy. It was a topic of many heated discussions, and I usually left the camera in the box when home. On this particular night, I was hoping to catch a glimpse of our intruder, but the thing was either too fast, smart, or lucky.

There were no vibrations. My gut wasn't sounding an alarm. There hadn't been contact or a blue-green light, but still I suspected this creature wasn't of our realm. It raced past our bed and through the closed glass door between our bedroom and backyard. The metal blinds rustled and moved as the creature passed through before making sounds outside. And while the rats in Toronto are pretty clever, they aren't magic. They can't run through doors.

I was tired of this creature robbing me of sleep. Lisa had woken a few times since it had started sprinting through our room. She'd mumble, "What was that?" then fall back to sleep when I assured her it was nothing. *Deny, deny, deny. See how easy this is?*

While Lisa knew of my abductions and late-night visitations, we didn't speak of them. I did a good job of gathering my experiences back into my shell where I could pretend that they didn't exist, and Lisa had a shell of her own. The odd time we did speak of creatures from other planets, Lisa was supportive but distant. How can a person relate to something they've never experienced?

I must have drifted off because I was startled awake by a strong vibration, and when I opened my eyes, I was floating in space. I was suspended in midair, surrounded by what appeared to be a galaxy. Actual outer space.

Once the initial shock wore off, fear set in. *Holy shit, I'm floating in space.* My heart was racing. My breathing came in spurts. I could see and feel my arms and legs, and I knew I was me, Robert Hunt, but the view looked real. There was no way this was a dream. I was still wearing my pajamas.

The reality of what was happening sunk in. I wasn't hot or cold. I wasn't exploding. I felt fine. I could breathe, and the air felt somewhat normal. I couldn't see or feel my bed or bedroom, and I could no longer hear Lisa sleeping beside me. There was this strange sensation around my head and face. It didn't hurt, and it had no substance, but I could feel it, like a cloud hovering around me. I ignored it to look around. After all, I was in space.

I turned in circles, taking in the sight. The vast galaxy before me was massive and seemed to go on forever. There were planets everywhere, more than I could count. Some were bright, reflecting light coming from other planets, gases, or suns, perhaps. Some were dark. They were all different sizes and colors, and the perspective was so strange I couldn't register what I was seeing. Huge brightly colored clouds seemed to hover around the planets. I couldn't see movement, and the depth and dimensions of the clouds were impossible to decipher. Still, they existed as blotches of colors I'd never thought possible.

Two planets felt close enough to touch. The one to the right appeared to be either closer or much larger than the left. It was big and bright. I could see the structure of the planet's surface. Unlike pictures I'd seen of earth from space, this planet was mostly green. There were white areas that looked like cloud cover and big blue oceans or bodies of water, but the rest was comprised of vibrant shades of green.

While the view was crystal clear, I couldn't get my head around what I was seeing. How was this possible? What was happening, and who or what was showing me this vision? There were no spaceships or beings. There were no voices or instructions. I couldn't see anything beyond the vast galaxy before me, yet I knew this was a vision of sorts. There was no other explanation.

To the left of the big planet was a smaller one. It was impossible to tell if it was smaller in size or so far away it only appeared small, but from my perspective, it looked very close and less than half the size of the planet beside it. This small planet was bluish, like earth. It was much harder to make out the surface, but I could see what looked like bodies of water and what appeared to be mountains covering the majority of the surface.

What really struck me was how the planets appeared to be turning in different directions. The large planet was turning slowly, yet enough to see movement. It was turning to the left—counter-clockwise. The smaller planet was spinning much faster and was turning to the right, clockwise.

In the blink of an eye, the galaxy was gone, and I was back in bed, wide awake, staring into the dark abyss of my bedroom. Lisa was gone, along with her bedside book. She must have woken up and gone to the living room to read. The cloud that had been hovering around my face had also disappeared, and I couldn't hear the rat-a-tat-tat of flat feet on the floors.

It had been an exhilarating experience. So much so, I find it hard to put into words. I know nothing of the stars or astronomy. What, exactly, did I see? Who or what showed me these planets, this galaxy? Why?

All I know is that a few days later, when I happened to check the trail camera that I'd hidden in hopes of catching a flat-footed rodent, I'd caught something altogether different. I was amazed to see what appeared to be a ball of light, energy, or something else floating beside my bed, then hovering around my face—like a cloud.

That cloud showed me something I will never forget. I only wish I understood the meaning or message.

Notice the halo around my head, like a cloud. The room is pitch dark and yet my eyes are open, looking into this hovering cloud.

21

NOVEMBER 2013

DIMENSIONS OF GRANDEUR

That same year, 2013, only several months later, I learned the hard way that extraterrestrial beings weren't the only out-there lifeforms interested in the human race.

At the time, I was still working in Oyen, Alberta, where I worked long hours and slept at a local hotel. I had the routine down like clockwork, and this night was no exception. By 9:30 p.m., I'd washed the day away, set up the trail camera, and was snuggled in bed with my pillow propped behind me and my tablet on my lap. I was watching a movie.

My right arm started to tingle. At first, I brushed the feeling off, thinking it was the sensation you get when you've fallen asleep on your arm and it's gone numb. I hadn't done that, but still. My gut wasn't telling me it was time for company, and the usual vibrations weren't present. The lights were on. I looked around the room. I was a guy in bed watching a movie.

The feeling didn't go away. It got stronger. My arm got heavy, as if made of lead.

I shifted in bed, wondering if I'd put pressure on a nerve or hit my arm without knowing. I held my heavy arm upright, searching for clues but saw nothing. Within minutes, the feeling turned from a tingle to a throb that didn't fade.

It was now almost painful. The muscles in my arm contracted and pulled, tightening in a way that scared the heck out of me, and, for a moment, I thought I was having another heart attack or a medical emergency. I was about to grab the phone to call 911 when this strange cloud formed around my right hand.

A ring of what resembled smoke circled my hand. It looked like someone had blown a thick ring of cigarette smoke at me. It was white, dense, and spinning around my hand. It made no sound. I don't recall an odor, and if the smoke had a texture or feel, I couldn't sense it over the pain. The cloud moved slowly, but the pain got more and more intense.

A searing pain hit me. I doubled over in bed and was about to scramble for the door when a being of some sort literally stepped out of my hand and onto my bunched-up knees. The circling smoke had made it hard to see what was going on, but it was as if this creature had stepped out of my flesh, separating itself from me, and formed on my lap.

I simply stared. What else could I do? I was looking at a creature that just came from my hand! It had a face. There were two eyes, a nose, and a mouth, positioned as on any human face, but surely not human. Its body was out of proportion—small on top and larger on the bottom—and it moved under some clothing or cover that was solid and dark. The lower part of this being didn't seem to have a beginning or an end. It formed from the quilt like it needed a source.

I didn't know what to make of this creature who stared at me with no expression I could recognize, no personality or purpose I could pinpoint. It just stood there, on me, with no more weight than a light-footed cat.

The pain had vanished. The heaviness in my arm and hand was gone. The cloud of smoke had disappeared. A wave of serenity washed over me. In an instant, I felt calm, as if none of this wild event was taking place. The creature and I looked into each other's eyes with utter simplicity. Neither of us said a word. I was suddenly happy, content, and I didn't mind the intrusion. This being wasn't here to cause harm. It wasn't trying to hurt me or cause discomfort. I had the implicit feeling I wasn't to fear this thing—as though it had spoken words of kinship.

I remember nothing past that moment.

If the creature returned to where it came from through my hand, I didn't suffer the tingling, weight, or pain I'd experienced on its arrival. If it stayed here, on earth, and left the hotel room through the door or wall, I didn't see it go. I have no idea how or why this being manifested through my body. I don't know what I was supposed to learn from the experience—if anything. It gave me no grand message or understanding.

I woke at exactly 2:07 a.m. The light was still on, and I was sitting in bed with the pillow propped behind me. My tablet was turned off at my side.

I didn't know it at the time, but this visitation would mark another major turning point in my life. That night, my encounters with solid, three-dimensional extraterrestrials would take a backseat to something altogether different and unbelievable.

I would soon learn the difference between extraterrestrial beings from other planets and extradimensional beings who originate and exist beyond our physical reality. That November night in 2013 was my first encounter with beings from a dimension beyond our own—beings from a lower or higher dimension.

Little did I know, it wouldn't be the last.

22

WILD THINGS

Things only got stranger. The floodgates were open, and beings without physical form flowed through. From the winter of 2014 onward, I was visited by creatures who didn't abduct me or communicate, but made themselves known through creative means.

Any gut instinct I had, including my internal clock, didn't function well around these beings. The vibration they gave off was slight—not debilitating or paralyzing like the Grays or others—yet it was there, and it was the only warning I received.

Like extraterrestrials, these creatures could find me anywhere. At a maple sugar shack during a short couple's getaway with Lisa. At hotels and motels across the country, where I slept when working away from home. At our house in Toronto. In the car.

Their visits felt random and I don't know what they wanted yet, when they were present, I was drawn to them, as if called or summoned through my subconscious. They formed faces, body parts, or hands—mostly hands—with the use of bedding, clothing, and objects they could manipulate. They knocked things over, pushed stuff off tables, and toppled what was upright. They pulled the sheets off my bed, poked me, and made sounds with doors, windows, and whatever got my attention.

While some of these events were creepy, most were mainly perplexing.

I preferred these visits over extraterrestrial abductions—who wouldn't? These creatures didn't steal me from my bed or probe me on an operating table like a piece of meat. Still, had I not experienced a lifetime of unexplained phenomena and encounters with non-human lifeforms, I would have thought myself crazy or delusional.

But I knew better. I knew the unexplained and unmentionable had substance.

I was also convinced these beings weren't messing with me alone. On more than one occasion, I caught my wife Lisa having conversations with faces formed on the pillow beside her—while unaware. Once, I even found her laughing, petting a creature she couldn't see, but could clearly interact with when she was supposed to be sleeping. She had no recollection of these instances. Like most spouses present during extraterrestrial experiences, she was somehow oblivious. It seemed aliens put a lot of effort into showing me they existed and how they project their consciousness into our reality. There had to be a rhyme or reason for it. There had to be.

While a large percentage of abductions take place at night, when a person is sleeping, this could be a matter of convenience. Experiencers are abducted from wherever they are, at any time of day or night. They've been taken from their homes, beds, cars, etc., and no geographical area seems out-of-bounds.

Is it random? Are people just plucked from their beds willy-nilly? I don't think so. Considering I've been visited and abducted dozens of times from pretty well anywhere I've been, coincidence is unlikely. They find me everywhere. And abductees are almost always returned to their bed or car, to the scene of the crime. How is this possible? How do aliens know where a person is at any given moment?

While I don't claim to know, plenty of abductees believe they are tracked with an implant—tagged like a wild animal. For some, this is a physical object they can feel or see within their body. Common areas are the head, eyes, nose, eyebrows, hands, or feet. These objects were not present before an abduction and cannot be traced to a medical procedure performed by a human doctor. Other experiencers don't find physical proof but feel a tracking device has been implanted unseen. It's a chicken-or-the-egg situation. Did an implant track an abductee, or was she found by other means, making her feel traced by a physical object? Who knows?

In 1995, Dr. David Prichard, a medical physician, reported removing implants of unknown origin from the toes of several abductees under strict scientific conditions. Analysis of the objects was inconclusive. Visually, the pieces resembled glass, metal, or ceramic, but the materials defied scientific classification. No one had a clue what these things were.

In the United States, Dr. Roger Leir removed objects believed to be alien-made tracking devices from patients. These objects ranged in shape and size, but most were tiny, dark metal pieces of unknown origin. He even wrote a book about his findings, titled *Alien Implants*.

Since most abductees, like me, can sense an impending alien visitation, it's not unrealistic to think tracking devices could have something to do with the heads-up. Again, I often feel subtle vibrations hours before an extraterrestrial event. These aren't normal vibrations like something from a piece of man-made equipment. These vibrations touch my entire body. They sort of feel like a muscle-relaxing machine on high. The vibrations are sometimes stronger or weaker, and the duration isn't always the same. Still, I never doubt the vibrations are connected to alien phenomena, even when they don't result in abductions I can recall. I just know, like knowing how to breathe. I can't explain how.

When I was a teenager, my best friend and I loved to wreak havoc in my backyard. We had a motorized mini dirt bike we'd zoom around the yard at full throttle. My dad was an avid moose hunter who used a heavy canvas tent on hunting trips. Before hunting season started, he'd set up the tent in the backyard to air out. Linda, my younger sister, used it for sleepovers with her friends, so it stayed for weeks on end, even when it was in the way of my fun.

One day, I hauled that sucker out of the way in my bare feet and shorts. "Too bad," I said when my sister ran inside to tattle. I hopped back on the bike without a care.

On the third round around the yard, my foot caught on a tent peg I'd missed. I fell off the bike, hollering in pain, rolling on the ground. Within minutes my foot swelled to twice its size. Dark blotches of bruising crept up the sole to my ankle. The skin between my big and second toe was torn open and the toenail had been ripped from the nailbed of my big toe. My parents weren't home. My uncle raced me to the local hospital, thinking I'd broken bones. That's how bad it looked.

An emergency doctor came into the room with x-rays of my foot. He glanced at my uncle before giving me the once-over. *Typical teenager.* He held the x-rays up to the light one at a time. "You're a lucky guy," he said. "No broken bones, and the pin isn't damaged."

Pin? What pin? I don't have any pins in my body. I also had no foot scars from a previous surgery or break.

"Had you broken the pin in your big toe," the doc continued, "you'd be staying for surgery." He tucked the x-rays into an orange envelope and turned to my uncle with instructions for my care. Confused, I only half-listened. Nothing had ever happened to my feet. I'd hardly ever been to a hospital, and I'd never had surgery to put a pin in my big toe.

At least, not that I knew of, and not by any human doctor.

Afterward, I never gave the pin much thought. Either the doctor had been looking at someone else's x-rays or he was getting my case confused with another one from that busy day. The obvious explanation was human error, a medical mix-up—they happen all the time. There was no pin. I've only recently come to admit, it could be an alien implant. Part of me wants to know the truth. A bigger part says hell no.

I don't feel like I've got anything foreign inside me. Extraterrestrials have never said anything to make me believe there is a tracking device or implant in my body, and I haven't witnessed an object being inserted, monitored, or maintained in any way. If there is a pin in my toe, and it's used to track my whereabouts, there's nothing to be done about it anyway. I'm sure that if I did have surgery to remove it, the creatures I've encountered would be able to find me easily enough without it. I learned to live with the cards I was dealt.

23

ANSWERS

The night of answers was a hot, humid one in August. I was crashing at an Ottawa hotel close to a job site. I had followed my usual routine: clean-up, dinner, a bit of television, and bed by ten o'clock.

Asleep for a while, I felt something moving. It was climbing up the bed beside me, pushing me in the process, and I shot up to a sitting position to look. A haze of light from the hotel parking lot shone into the room, and I caught a glimpse of what looked like two dog-like creatures making their way up my left side. I would later learn that many experiencers encounter creatures that resemble animals, reptiles, and even insects. But at this moment, I had never heard of such things. I was petrified. I struggled to jump out of bed but, as usual, I was pinned.

A loud humming permeated the room and something emerged between the dogs and pushed toward me. This thing was tubular, with no legs. It moved like a snake. I tried to cry out but, in this instance, too, I was muted. The creature was big, much larger than the dogs or any snake I'd ever seen. The buzzing got louder as the creature reared before me with its snake-like face and dark spots for eyes. I was in shock.

The room went black, as suddenly as if someone had pulled a hat over my eyes. All I could see was a faint shimmer of movement in the lowest range of my vision. There was music—a waltz. The snake and the dogs were gone. I was suddenly spinning in the pitch dark with someone holding my hands. Their fingers were entwined with mine as we turned around and around, dancing to the music. There was a powerful vibration but no pain, and the second I registered my feet on solid ground, not moving with the rest of my body, I pulled my hands away. The music stopped. *Where was I? Who or what was in the dark with me?*

It never entered my mind to speak. I'd had a lifetime of nonverbal interactions. I reached out, groping the black abyss that surrounded me like a blanket. There was no sound, no smell, nothing. My fingertips touched something in the dark, and on instinct, I stepped closer. I don't know why, but I wasn't afraid. I could feel what felt like shoulders, soft skin, and hair—human hair. This creature was shorter than me. I felt my way through a full head of long, straight hair, two eyes, a nose, and a mouth. This was a normal face, the face of a person, and I was confused.

It didn't say a word as I felt my way down its neck and chest, gasping as my hands touched familiar ground, something like female breasts, and I pulled my hands away. "You're a girl," I cried out, surprised that I could speak.

A laugh filled the darkness. She sounded young, no older than my daughter. "I am a girl, silly," she said. Her voice was fluid. She spoke like a local.

Questions pushed and shoved their way to the forefront of my mind but I didn't get the chance to ask them. She grabbed my hand and pulled me out of the darkness into what looked like an old living room. The vibration here was so powerful I could hardly remain upright. My body went heavy, like I was suddenly 400 pounds. I felt very ill, and while I tried to gain my equilibrium, the girl led me to the middle of the room and lowered me to the floor. A large window flooded the room with sunlight. I could feel the heat on my skin. The living room was sparse, with only a sofa flagged by an end table and a coffee table. There was one chair in the corner by the window.

I was having trouble breathing. I wasn't sure I could remain upright, so I braced myself and tried to look around the room. There wasn't a single picture or knick-knack. The furniture had been pulled from the sixties. There was no television. Although barren, this place looked human, like the apartment of a student. A black box sat on an end table. It was rectangular and didn't look like anything I recognized—a stereo maybe? There were no knobs or controls, only a small round light on the front.

Different music started playing. The waltz was over. This sounded more like soft rock.

There were stairs to my right, heading down. When I looked behind me, where we came from, there was nothing, just black space, as if the living room just ended. I can't explain it, but there was no door or wall, no *nothing*. Yet I was sure we came from that area of darkness, so it had to be something.

The girl sat in front of me on the floor, crossing her legs and spreading her sleeveless white summer dress around her. The window was at her back, the sun's rays framing her like a halo. She was definitely human—or human-looking. She couldn't have been more than twenty-two. Her skin color was like mine, Caucasian, and her long dark hair draped over her shoulders. She was pretty. Her smile calmed me.

She spoke, but I couldn't hear her. My body was shuddering with vibrations. Everything ached. I was having trouble holding my head up, and the music was loud and overwhelming. Someone entered the room from behind me, walking around the girl's back before joining us on the floor. He leaned over to say something to the girl, and she smiled. He was also young, maybe younger than her. He had short dark hair, clean cut. He was dressed in something casual and neutral. I remember he was wearing white socks with no shoes.

"You were brought here . . ." The girl was speaking to me, but I could only make out a few words. It took everything in me to say, "I can't hear you."

She looked over my left shoulder and squinted toward the black box on the end table. The music lowered, suddenly tolerable.

"Robert," she said, "the dance was to welcome you to this higher level of vibrational consciousness. You know this as Ascension."

She said my name!

I had no idea what a higher level of vibrational consciousness was, but I tried my damnedest to listen and remember what she was saying. As ill as I felt, there was something in me that knew this was important.

"The dance was a grand gesture," she said. "Welcome to this higher vibrational reality. This is your first stage of Ascension." She smiled, possibly trying to reassure me. *What is a higher level of reality? Would I survive it? What the hell was Ascension?*

"You've encountered many entities who helped you get to this stage," she said. "They were preparing you for this higher level of consciousness."

Snippets of questions floated through my mind—*Preparing me? Who—which creatures? What did they do?*—but none made it to my mouth. I could hardly understand what she was saying. In this state, my mind wasn't capable of analyzing facts. It took all my concentration to absorb what she was telling me.

She leaned closer. "This will be your new home," she said.

I assumed she meant this place outside the window, not this apartment, but how could I know? Was this a higher vibrational place, a world within our world?

"When you return," she said, "you will meet a man with a long white beard and robe. He will be wearing sandals. He will give you a tampon."

This, of all things, struck me. A tampon?

Why would someone give me a tampon? "Am I going to be a girl?" It was the only question I could spit out.

She exchanged a look with the guy beside her but didn't answer me. "Robert, I have to go now," she said. She rolled forward and hugged me with one arm. She felt solid, real. "You'll be fine."

I watched her get to her feet and leave down the stairs.

"Come on," said the young man, rising from the floor. "I'm going to stand you up so you can see out the window."

He, too, was solid, all muscle and bone. He helped me get up and shuffle to the window. It was bright outside, although I don't remember seeing the sun. I could feel heat on my face. The warmth of the window ledge radiated through my hand. The sky was blue and cloudless. We were one or two stories up. Green grass and a long row of mature maple trees were on the other side of the paved road. The branches were swaying in the slight breeze. The trees were large, taller than the building I was in, and I remember noticing the leaves, Maple leaves, and thinking of home. There was a black asphalt road in front of the building, but no cars. A concrete sidewalk lined the close side of the street. I tried to focus on the lines in the concrete.

Despite the familiar objects I was seeing, there was something strange about the view. It was as if it was missing detail, but I couldn't put my finger on it. I was so sore, so tired.

After some time, I noticed the girl walking down the sidewalk. She was the only person outside. Her white dress was dancing with the wind. She stopped, looked over her shoulder, and then up to the window where I was looking out. She smiled and turned away, back on her path to somewhere.

"It's time to go home," said the guy behind me. I'd forgotten he was there. He led me to the dark area near the far end of the living room. I could hardly walk. The vibrations were killing me. I don't remember him saying goodbye. I only know I woke in the hotel room, sitting up in bed, sick as a dog.

My head was spinning. I sat there, replaying what had happened in my mind. I grabbed the pad of paper and the hotel pen from the bedside table drawer. I had to write things down before the details escaped me. I had to remember what the girl had told me. There was nothing more important. I had to record it, analyze it, and understand it. She said: "The dance was to welcome you to this higher level of vibrational consciousness." There is another place, possibly another dimension, that exists on a different vibrational level. I was there, but my body wasn't ready for it. The higher vibration made me very ill. It looked like earth, like home, only different. There might be people or creatures there who look and feel like I do. To get there, I have to ascend.

The girl had said, "You know this as Ascension." By you, I assumed she meant human beings in general. What did mankind know about Ascension, and what did aliens know it as? The only thing I knew about Ascension was the general dictionary meaning: to rise or move up to a higher position. I had a faint recollection of a Christian or Catholic holiday centered around Jesus dying and rising to heaven, but since I wasn't religious, that was all I knew.

"You've encountered many entities who helped you get to this stage," she'd said. "They were preparing you for this higher level of consciousness." So, there was a link, a purpose, a reason I'd been abducted all those years. They were trying to help me get to this higher level of consciousness. By doing what to me? I didn't know, but the glass rod and classroom came to mind.

"This will be your new home." I was heading back there and staying forever. Man, I hoped this next level of consciousness wasn't in the sixties.

"When you return," she'd said, "you will meet a man with a long white beard and robe. He will be wearing sandals." Someone was going to greet me there, a man. This man sounded an awful lot like the typical description of a religious figure. Jesus? God? Was there a reason for the meeting? I didn't get the impression this experience had a religious purpose.

"He will give you a tampon." This guy will give me a tampon. A tampon! Does that mean I'll be female in this other dimension? If

I'm a woman there, does that mean my body here will be dead? Will I be reincarnated? Will the soul or energy within me move to the body of a girl, or will I continue here at the same time? I didn't fear death. I knew we all had to die at some point. Still, I didn't want to leave my loved ones. Would my loved ones come to this higher dimension with me? Would everyone—all of mankind—be rising to this higher place?

When would this happen?

The minute I thought this, a voice chimed in my head, loud and uncomplicated. "One Sunday in August," it said.

This wasn't the voice of my consciousness, or the girl, or the guy from the apartment. This was a deep and unfamiliar man's voice. There was no accent, and his words couldn't have been clearer.

I didn't have the wits about me to ask another question. Where would I start? I was already reeling from the experience. Something was going to happen to me—something earth-shatteringly big—one Sunday in August. He didn't say what year, just one Sunday in August.

That day, August 16, 2017, was the first time I ever recorded an experience on paper.

I was a survivor, an experiencer. And after an entire lifetime of abductions and supernatural encounters, I had finally met something or someone who left me with more than just questions.

For once, I got answers.

Then, something appeared. I don't know where I went, or who was with me. Everything happened at 2:47 a.m—in a matter of seconds.

Here, I'm missing from the room entirely. Notice it is still 2:47 a.m. Thirteen minutes later, at 3:00 a.m., I'm returned to my bed.

24

AUGUST 19, 2017

EXTRADIMENSIONAL FANFARE

Having answers, clues to understanding why I'd experienced a lifetime of craziness, was freeing. There was now meaning to my life—reasons for the extraterrestrial interactions and strange phenomena.

Like a madman, I dove into researching the concepts the young woman in the apartment had told me about. I was determined to learn about higher vibrational levels, possible vibrational dimensions, and Ascension.

I was obsessed. Why wouldn't I be? Some future Sunday in August would change my life. It was like knowing when you'd die and understanding your life wouldn't end when you did. You'd go on to live another day, week, year . . . another life.

I returned home after that experience in Ottawa. I couldn't concentrate at work. I needed to be surrounded by familiar things. The kids had spread their wings, landing in new but exciting places. Lisa was busy with her work. While I knew home was where I had to be at the time, I was having trouble relaxing and sleeping. My mind was on overdrive. There was so much to learn. Technology had advanced by leaps and bounds since the years of my childhood, when help was only found from a library book, friend, or family member. The Internet presented me with an endless sea of information, advice, and perspective. Weeding through the shallows was the hard part. The line between fact and fiction was blurred—if it existed at all. Religious sites shared concepts and perspectives through the lens of belief. Government opinion was obscured by politics or shrouded in conspiracy theories. Even science was open to interpretation. Learning wasn't an easy process. It still isn't.

I found myself fascinated with vibrational levels and how scientists believe they've discovered a universal signature of life that takes the form of vibrations given out by all living things. We are all made of matter, and they say matter is constantly in motion, even when visibly stationary. Every living creature on our planet emits tiny vibrations with a distinctive hum. Researchers have recorded this vibration in animal, human, and plant cells, as well as rudimentary lifeforms such as bacteria, yeast, and cancer cells. We are all just energy resonating at a certain frequency. And when matter comes together, close to other matter, it begins to sync up, to vibrate at the same or similar frequency. It forms a connection.

Scientists have also discovered that our vibrations are linked to our consciousness. In other words, our vibrational level is more than our physical components. Some refer to this as our spiritual vibration. In basic terms, negative thought processes such as anger, fear, jealousy, and hatred emit low vibrational frequencies. Positive emotions such as love, gratitude, kindness, and caring create high vibrational frequencies. When you, as matter, resonate at a certain frequency, you attract other matter with the same or similar vibration level, connecting for the greater good.

My mind blown, this led me to think about the vibrations I'd been racked with when otherworldly creatures were close. Was I feeling their vibrational output, their signature of life? Or was I feeling the vibrational level from where they exist? Were these concepts the same? And what about the debilitating vibration that hit me in the apartment with the young woman and man, where I felt so sick? Had I felt the vibrations given off by the living things in that higher dimension, their different vibrational frequency level? Since nature, our natural environment, is alive, does every creature exist within a specific vibrational level? Is traveling or living within one or the other only possible when one's inner or spiritual vibration is compatible?

Was this why different lifeforms, both extraterrestrial and extradimensional, radiated with unique levels of vibration—a difference I'd felt my whole life, since I was a boy of seven? The possibilities were endless.

For nights I'd been tossing and turning in bed, thoughts shooting around my head like a pinball. On the night of August 19, only three nights after I'd been visited by the girl and boy from another dimension, sleep was fraught with conflicting thoughts and confusion. I was sure I'd been given the keys to my future, but I didn't know how they worked or what they meant. While access to information and opinions worldwide opened my eyes to new concepts I'd never considered, the sheer scope was overwhelming. How could I decipher the pertinent details?

I propped myself up in bed, my vision drawn to a spot of light hovering around the middle drawer of the dresser in our room. The bedroom was mostly dark. Lisa was sleeping soundly beside me. I watched the light as it flickered—not really thinking it was anything usual. I was grateful for the distraction.

Until the light started to grow.

It grew larger and larger, engulfing the dresser with light—then the wall, the ceiling, and my entire range of vision. There was no edge, no start, and no end. It was as if I was outside amidst a giant movie screen.

The light morphed into form. Things were moving toward me.

They first appeared as stickmen, but as they came closer, I realized they were cartoon-like toy soldiers or old tin soldiers. In bright-colored uniforms, they marched in three rows, at least a dozen or two in length. They weren't real. They were animations.

I don't recall if I was physically present, but I know my perspective was street-level as the show came into view. The entire setting was animated. Surrounding me were buildings drawn in bright colors with very little detail. Like row houses, they were continuous and ran down each side of a wide street lined with streetlights. The sky beyond was light blue. Before me—a good 200 feet in front of the soldiers—were three cannons on wheels. They were tilted upward toward the sky. These were also cartoonish. One was yellow, one red, and one blue.

I don't recall hearing music or sound, but the scene reminded me of a parade or celebration.

I was standing behind the center cannon, the red one, as the first soldier from each line ran up to the cannons. They paused for a moment, looking in my direction, then pulled what looked like chains on the sides of the artillery.

There was a loud boom as three black cannonballs shot into the air. The sound was thunderous, and my first reaction was to look for Lisa, to see if the sound had scared her, but she was no longer beside me. Our bed, our bedroom, was gone, and I was watching the show alone.

I watched the cannonballs rise into the air surrounded by animated stars, like when things blow up in cartoons. The balls hovered in midair for a moment, then exploded into a big yellow blast accompanied by more stars. A blue bird flew from each blast, soaring into the sky above the three rows of soldiers. I could hear their wings flapping in the air.

This happened over and over, the first soldier in each line igniting the cannons and cannonballs exploding to reveal birds. I'd never seen anything like it. I was mesmerized.

I knew the show was coming to an end when the scene started to fade out. Darkness closed in around me until all that was left was pitch black. If this was a hallucination, I must have been on some mighty fine drugs. The show was incredible, beyond anything my imagination could concoct.

Then a small light appeared, hovering in front of me, similar to the flicker I'd seen before the show. It expanded until a creature's head appeared—only the head. This being was not animated. It looked familiar. This was the same face I'd seen within my bedsheets on various occasions, the extradimensional being with eyes, a nose, and a mouth. Its forehead and chin protruded outward. Something about it made me think it was male. When he moved, turning slightly to one side, his profile exposed a bone or fin jutting from the back of his head. He looked almost aquatic. Either the creature was pale gray or I was seeing in black-and-white. He smiled at me. He had a fantastic smile—I have no other way to describe it. He radiated kindness and sincerity.

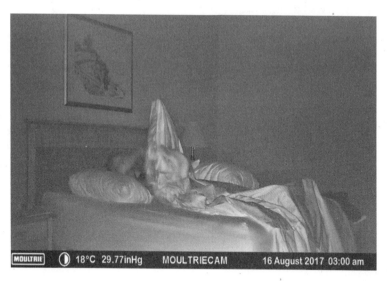

This is a picture from the previous chapter, dated August 16, 2017. After the extradimensional fanfare on August 19, I was looking through pictures and spotted the head of the creature I saw at the end of the fanfare celebration— the movie. So he was present the night I was taken to the apartment, although I didn't know it at the time. Notice his chin and forehead and the shaded fin protruding from the back of his head.

He mouthed the words *I love you*. My reaction was instant, instinctual. I did the same. *I love you*. This wasn't love in any romantic sense but a love shared by two species—respect for another life. He smiled again.

As the light faded out, my bedroom came back into focus. Lisa was sleeping in bed beside me, snoring soundly. The room was dark—the light was gone. I felt tingly, energized, like I'd just climbed into bed after a major party.

Whoever these beings were, wherever they were from, they wanted desperately to show me that my existence was celebrated. The show felt like a ceremony of some sort, an event meant to mark my connection to something much bigger than me. I'd lived a life beyond the norm—a fantastic journey. Along with the dance with the female from the higher dimension, this show was a welcoming, a spiritual awakening. I was part of something bigger than words could ever describe.

I felt it.

25

MULTIPLE REALITIES

I was on the track to enlightenment. There was so much to learn, to understand. But I wasn't alone. I had teachers. Only a few months after I'd received answers and was celebrated, they showed me more.

I woke somewhere I'd never been, in front of people I'd never met. A woman was standing in the hallway of what I assumed was her home. Behind her was the kitchen. She looked to be in her mid-eighties. Her eyes pierced mine as she stood there, staring at me.

A younger woman passed me in the hallway. She was carrying bags of groceries and didn't seem to notice me. "What are you staring at, Mom?" she said as she turned sideways to squeeze past the older woman. The groceries grumbled as the woman plopped the bags on the counter beside the fridge.

I stood there, frozen.

The older woman wouldn't take her eyes off me. She wasn't afraid. She radiated a sense of sadness I could feel as if it were an emotion I could touch. Her loss was profound. I'd never felt such a heavy feeling. It consumed me. She kept staring into my eyes.

"It will be okay," I said. I don't know if I spoke or thought the words, but I know I reacted on impulse. It felt right.

The woman smiled at me, emotion filling her eyes. Her shoulders relaxed. She understood.

In a blink I was somewhere else, as if lifted and placed.

To my right was a Great Dane sitting on the end of a couch. I suppose I should've been afraid of the dog, but I wasn't. He made no move toward me. He merely looked at me, head cocked. He was panting as if he'd just come in from a walk. An early-morning sun shone through the window. We were high up—in an apartment, maybe? I could see the tops of trees. To my left, a woman came out of what looked like a bathroom.

I think my jaw dropped. It was my mother-in-law, Georgia!

I couldn't believe it. Georgia, at the time, was in her late seventies. Here, she wasn't more than forty or forty-five. She was dressed in a nice pantsuit with her hair done.

"There are no more towels," she yelled.

Even her voice sounded younger, vibrant. The dog noticed my confusion, I think, but Georgia didn't. She didn't see me at all. She scowled at the dog.

A voice responded from somewhere in the apartment. "Bring something, anything, but hurry." I recognized the voice. It was Georgia's daughter, Nicky, my sister-in-law. "The dog pissed everywhere!"

It looked as though Georgia and Nicky lived together in this apartment. *But this can't be*, I thought. They didn't live together, not since Nicky was a kid. They now lived in different cities and had hardly talked to each other in forty years.

Once again, the lifting feeling came over me.

I was suddenly standing beside a young girl in a hardware store. She was twentyish, maybe twenty-five. She was searching for something. She looked at me, not startled that I'd appeared beside her from out of nowhere.

"What are you doing?" I asked.

"I'm trying to find something," she said. "It might be on top." She pointed to the top of the tall shelving unit.

I took her right hand. I was shocked I had done such a thing, but the girl didn't even flinch. "I'll help you," I said.

I have no idea how, but the two of us floated off our feet, into the air, and over the aisle. My heart was beating outside my chest. I watched as the girl conducted her search, never once showing signs of distress. She found what she was looking for, and we returned to the ground, landing on our feet next to an in-store restaurant.

"I think I'll have an early lunch," the girl said, smiling at me. She turned and presumably went for lunch. And that was it.

Now what do I do?

I headed to the front of the store. I wanted to see outside, to know where I was. It appeared I didn't need my feet: I was floating again. It was the strangest sensation, as if I could move with my mind. I went through the wall to the right of the exit doors as if this was a normal thing to do. *I can float through walls.* Only, as I turned around, bricks were falling out of the wall onto the sidewalk. *Shit, I can't float through walls.*

A man approached me, ushering me away from the hardware shop, guiding me toward an adjacent grocery store. I followed without hesitation—I'd ruined a brick wall. As we entered the grocery store, two ladies stopped to stare at me. One leaned over her grocery cart to whisper to the other. "Is that him?" she said.

Me? Why all the fuss?

Confused, I followed the man to the back of the store. My memory is hazy from there, but I think we exited the store and got into some sort of vehicle.

Another blink and my reality changed in an instant.

I was falling through the roof of a house. I could see the roof below me, then over me, and my feet were planted firmly on a floor. I'd been plopped in the middle of a living room while a family sat on a sofa watching television. The couch was quite long—too long. A teenage girl and a woman I assumed was her mother pulled their gaze from the old, bulky-looking television. They were startled to see me.

"What the!" the man said, jumping from the sofa. He looked furious. He was a foot away from me when I raised my hands in defense.

"I am lost," I said.

He stopped. A grin spread across his face.

"Oh," said the woman. She and her daughter got up from the couch. "Let me show you the way home." She took my hand. Her skin was warm and human.

I remember walking out the front door of their home and turning left on the sidewalk, following the woman down the street. The houses were old, most had stone fronts, with modern windows and doors. It resembled home, but it wasn't home. The details were familiar, yet not familiar. No words were spoken as we passed house after house. I was in awe.

After a few blocks, the lady stopped and pointed up, calling my attention to the sky.

What is this?

I'd never seen such a thing. People were floating or flying toward a massive black hole. It was round and hundreds of feet in diameter. I couldn't see anything in it, only pitch-black darkness. Upon closer inspection, I noticed the people were of various ages, coming from all directions. They weren't just leaving through the black hole. Some were arriving, appearing from the blackness.

I was spellbound. *How could this be?*

The woman squeezed my hand and smiled at me. *This was my way home.*

"Thank you," I said. "I appreciate you and your family's kindness."

I glanced at the big black hole in the sky before my feet left the ground. I recall moving, gliding through the air toward the portal. I wanted desperately to see something, anything within the portal, so

I knew where I was going. But there was only darkness. I followed the people. As I got close, I realized I was not afraid. My chest was brimming with excitement.

The next moment I was in my bedroom, still in bed, sitting up. Lisa was sound asleep beside me. My mind was spinning. In a matter of one evening, I'd traveled unknown dimensions. Time as I knew it had been shattered. I'd been shown multiple realities, places other than the world I know, yet the same, with human beings—even people I knew—existing in each and every one.

How could I see my mother-in-law exist in a place so different from home, from our reality? Why was she much younger? This wasn't the life I knew for her or Nicky, yet it seemed exactly that—another life. Who were the women in the grocery store, and why did they seem to know me? Why did the houses look the same but older, different? How did some people see me while others didn't? So many questions rushed through my head.

Could there be a link between time and dimension? If so, how could this be? Or, is it possible our concept of time is . . . off? For millennia, man's basic unit of time was daylight, monitored by the rising and setting of the sun. The measurement of time started with sundials, sandglasses, waterclocks, and candles, which were largely used to measure the duration of a specific activity. Even then, the vast majority of the population didn't have access to these means of measurement, making the sun's position their only option. Time was invented to organize currencies and international trade. Time is, by all accounts, a manufactured system that has very little to do with a clock. A clock merely allows us to track a day in twelve-hour periods, a.m. and p.m., dividing each hour into sixty minutes—all quite recent innovations in human history.

Time is, in fact, an illusion.

What if that illusion isn't linear? What if time isn't yesterday, today, or tomorrow, in chronological order? What if we have access to places and dimensions that exist within any point in time, through our conscious mind? What if we do this without knowing, living simultaneous versions of ourselves? Maybe every decision we make leads us to a different timeline or dimension? Or, perhaps we've already experienced various times and dimensions and are constantly reliving them, like a mass do-over. What if we have souls that experience consciousness in unique places and times at once, and each life is only limited to reality by a physical form like the human body?

There's so much to think about. Even now, when dwelling on the things I experienced that night, I have a hard time believing or understanding what I saw. I have no doubt the experiences were real. They were nothing like dreams or nightmares. I felt concrete, alive. The people and animals around me were living and breathing. All my senses were on overdrive. I know, deep down, these experiences were not random. They were shown to me by design, each connected to the next. They were lessons meant to show me—teach me—a new level of consciousness.

I suppose it's a good thing we cannot jump into other dimensions or times knowingly. Imagine the chaos. How would we cope? I'm not sure I could rationalize knowing, for certain, that I live several lives—possibly an infinite number. What would seeing and feeling this do to me? How would this knowledge affect my decisions?

Perhaps I'll find out someday.

I'm not the only one to experience phenomena that makes me question time and place. Many abductees claim they've experienced similar episodes. Some have even been shown their own lives in different dimensions. Dolores Cannon, a pioneer in the field of past-life regression, spent fifty years studying this concept at length, noting, "... time does not really exist as man identifies it. Every moment is now. The past, present, and future only exist in the now. We have been trained to view time as a linear progression of events based on the earth's rotation around the sun. Time is simply the perspective we adopt, and mankind is the only species in history to invent a way of measuring something that does not exist. Mankind will never truly reach the stars until we release our deeply ingrained concept of time and recognize the universal reality that everything exists in the now."

When I'm shown things, when I see things that are obviously not of this world as we know it, I have no concept of time. The sun is usually shining, but I have no idea what time of day it is. I don't know if my abductors track time as humans do or if they even age at the same rate as us. For all I know, they come from planets with other gravitational pulls and distant suns or light sources. Man adopted time to create structure here on earth, but I have no reason to think our version of time exists outside our planet as a linear invention.

While I know that what I saw, heard, and felt was unmistakable, there is still a part of my brain that screams no, this couldn't be; how could any of this be possible? How could I believe something that defies all logic, something beyond comprehension?

I was there.

26

YESTERDAY, TODAY, AND TOMORROW

SOS

"There are at least two trillion galaxies and twenty billion planets capable of supporting life. This is what we know, and these numbers could be infinitely more. There is no way our planet is the only one with life. In fact, there should be civilizations ahead of us. The galaxy has had over thirty billion years to create life and develop civilizations. Look what we've done in five hundred years. Imagine what other lifeforms could have done in millions of years."

BRIAN COX, PARTICLE PHYSICIST

Thousands of UFO sightings are reported every year across Canada and the United States alone. Other countries have similar numbers. But studies show these are only a small fraction of the sightings that actually take place. People don't report alien phenomena for various reasons.

Where would one even report such a thing? There are very few avenues to report unidentified flying objects or creatures not of this world, and none are openly shared with the public. There is no marketing campaign to inform or assist the general population with alien phenomena, no hotline, and no government info package. Even when someone has proof that an unexplainable event has occurred, science cannot determine what exactly happened and often the evidence is ignored to support the party line: aliens don't exist.

Yet they do. A simple online search will reveal survey after survey showing the vast majority of the world's population believes we are not alone—that other lifeforms exist.

Extraterrestrial beings visit and abduct humankind within every geographical location on earth, across borders and landscapes, from every walk of life. Religious beliefs hold no significance. Both men and women of all ages are taken. Skin color is moot. Political stance means nothing. You've got money—so what? It would seem these differences are inconsequential outside humanity's little bubble. Aliens are curious about our species as a whole.

So why are alien phenomena so hush-hush?

Just look around you. Our doctors and scientists who study the wonders of our planet and beyond are exceedingly cautious about exploration, and theories venture outside their proverbial boxes. Funding is hard to come by and is seldom granted to those who don't follow the current scientific dogma. We know every single lifeform on earth is made from the galaxy. Our ingredients were created by dead or dying stars, planets, connecting every one of us to the infinite universe. This has been proven by science, yet how many people are willing to face the implications of this knowledge? Only the daring or the financially independent.

Our governments have played a hand in the extreme discrimination against experiencers. Instead of altering the population's mass ignorance on the subject matter, they've created government arms like Project Blue Book, run in the United States from 1952 to 1969, to debunk credible sources. During a Senate hearing, Air Force Lt. Colonel Richard French, a major pillar within Project Blue Book, testified that his job was to silence UFO reports. He did his job well, although he eventually couldn't deny what he had seen.

"People who insist they've seen a UFO are nuts," say the critics.

"Abductees have psychological problems or sleep paralysis," claims the medical establishment.

"It's just astral projection, natural phenomena, or misinterpretation," insist the boxed scientists.

"What we know is classified," says the government, "but you're insane if you think we know anything."

"Experiencers are delusional!" screams the public.

Would you speak out, knowing the response you could expect? Is this a safe environment in which to share your unique experiences? How willing would you be to cut yourself open for the world to dissect?

The list of famous abductees discredited and broken by society is long and sad. Barney and Betty Hill spent their entire adult lives facing ridicule over the night in 1961 when the Grays took them from their car. Betty retold her story to the grave, never caving to the immense pressure to change her story for nonbelievers. Whitley Strieber wrote several books describing the phenomena he'd experienced and endured at the hands of the Grays. Within any other subject matter, the indecencies inflicted upon him as an abduction victim would be treated as such. Instead, he was shunned, laughed at, and debased by the media. Travis Walton, perhaps one of the most talked about abductees in the Western hemisphere, has spent the better part of his life telling his story. If you ever listen to or watch him recount those days in 1975 when aliens took him into an unidentified spacecraft, you would see how hard it is for him to think and speak of those memories, including the debilitating weeks that followed. The same could be said for the seven men who watched Travis disappear into the spaceship. Even after repeated interrogations and passing lie detector tests, these men were treated poorly for what they'd witnessed.

And it doesn't end with our systems, the media, or the public. While some are lucky to have supportive family and friends, many people who experience alien abduction or visitation are belittled by the family and friends they confide in.

No one wants to be famous for being stolen and probed by aliens. It's a humbling and sometimes humiliating experience, to say the least.

To understand the mind of an experiencer, you first need to consider the fact that most are abducted or visited early in life, when the experiencer is only a toddler or child. While alien visitations and abductions at this age are seldom violent, you can be sure these children suffer trauma.

Don't believe me? Imagine your child being awakened in the middle of the night by a strange hum or vibration and an unfamiliar light source that has invaded the sanctity of her bedroom. She feels and sees a creature unlike anything she's ever seen before—knowing this intruder should not be present. Screams and cries go unheard. It's clear she is alone and vulnerable, without you or her siblings to protect her. This is not a nightmare or her imagination. All senses are on overdrive as these creatures strip her naked and take her from her bed by force. There is limited struggle—she is physically powerless as she's pulled through walls, doors, windows, and ceilings by this beam of light. If she's able to see, her home and everything she knows recedes behind her as she's taken into an object of metal and light. Inside, an array of alien beings show her no warmth. They work her over in a surgical manner with instruments that look scary and cause distress and sometimes pain. The aliens take tissue samples, probe her face and head, and maybe enter her abdomen, anus, and reproductive organs. Fear is paramount. She is helpless and confused. Even if her physical scars disappear, the mental scars will not.

This alone, if experienced only once, would traumatize your child in immeasurable ways. Now imagine this happening over and over for months or years.

Can you envision such a thing?

It gets worse. The trauma doesn't end when your child is returned to her bed. When she runs to you for comfort and reassurance, what is she faced with? How do you, as a parent, react?

"It was just a dream, silly," you say. Maybe you take a quick peek. "I don't see anything on your body, so you must be fine."

Your daughter insists the experience was real, true.

"You've got an overactive imagination," your significant other pipes in, dismissing further inquiry. "Go back to bed."

These reactions, while understandable, given our human beliefs and biases, have an immense impact. Do kids have fantastical imaginations? Sure they do. Am I suggesting that every kid who has a nightmare about aliens really experienced the events they recite? Of course not. But for experiencers, those kids who have been abducted by aliens, the reaction of those around them contributes greatly to the trauma.

Even within the throes of an abduction, I struggled to grapple with the differences between reality and imagination when I was a kid. *This can't be real because my family told me it's not.* At seven, it didn't occur to me that my parents, the people I trusted to keep me safe, couldn't understand something they had never experienced. When my sisters taunted me, calling me names and making fun of the procedures I confessed I'd endured, I thought they were right and I was wrong. The reactions of adults, peers, mentors, and family profoundly affect a child's ability to process what they've seen and felt.

Emotions are kept hidden or buried in an effort to cope with experiences too profound for an undeveloped mind. Trust is lost. There is no such thing as a safe place when you can be taken at anytime, anywhere, and no one will save you. Nightmares and phobias disrupt sleep. The constant state of alertness can lead to eating problems and mood swings. Memories are sent to the nether regions of the mind as self-defense.

At the very least, these children learn to keep their mouths shut.

I certainly did. And this didn't change as I got older.

The more time passed, the more isolated I felt. Developing my self-identity was undoubtedly a different adventure as a teenager than most kids. My experiences didn't fit within mainstream belief systems or structures. The only reference I had involved jokes or negative stereotypes, so I deployed coping mechanisms to hide my secret life. Add hormones and major milestones like relationships and independence, and clearly being a teen or twenty-something abductee was not cool.

When I finally shed my reckless adolescence—thanks to Lisa—the risks of speaking out about my extraterrestrial experiences quadrupled. I'd gathered responsibilities. I now had a smart wife with big dreams of being a lawyer. We wanted a home. We'd developed friendships. We eventually had Josh and Keira, children who needed to eat. I got a real job, a career, and bosses who didn't consider experience with aliens a professional asset.

The irony was not lost on me. I'd climbed the corporate ladder to make a living and establish myself as a credible member of society. I was a reputable master electrician at Canada's largest nuclear power plants. But telling anyone almost anything about my alien encounters could result in everything I valued—my wife, kids, friends, our home, my reputation, my career, my ability to provide for my family—being taken from me.

Again, look around you. What do our systems and society do to people who speak out about alien phenomena? Experiencers are publicly shamed, ridiculed, and rejected. They are discarded as liars or fakes.

This is why people don't speak out. This is why UFOs and aliens are seldom discussed.

Even now, as a man in my sixties, sharing my secret is not easy. Just writing about my experiences makes me feel exposed in a way I am uncomfortable with.

But here's the rub: knowing aliens exist is a huge weight to carry.

When I was in my mid-fifties, I stumbled upon regression therapy. I'd never heard of such a thing—a doctor guiding a patient to remember past lives or memories buried deep. But the Internet opened my eyes to options I'd never considered and I was desperate to release decades of suppression. I learned there were different regression techniques to uncover memories the mind blocks out. I also discovered the doctors practicing these techniques on abduction patients were few and far between. Although it's true that the vast majority of the world's population believes alien lifeforms exist, very few certified professionals are willing to risk their reputation by admitting this. The ones willing to participate often subject their patients to harsh medical tests, medication, and hospitalization. Some abductees spend years with psychologists or counselors who cannot accept their patient's reality. Some experiencers seek medical advice in hopes of being diagnosed with a mental disorder—something peggable and treatable. No one wants to be right about abductions being real.

Still, I was intrigued by the possibility I could uncover more than my mind or my abductors allowed.

I found a psychologist practicing in Bowmanville, a reasonable distance from my home in Newcastle. While her website didn't mention alien abduction patients, she was an expert in regression therapy. With trepidation, I called to schedule an appointment.

Her office was small but pleasant. She asked me to relax on a leather sofa-lounger, and she sat in a chair she pulled close. I was nervous. In fact, I was petrified to confess all to a stranger. *Would she believe me? Would she think me insane?* At $350.00 an hour, I revealed only the basics of my life's unique story. I told her I'd been abducted by extraterrestrial beings my entire life, since the age of seven, and how I suspected a percentage of these experiences were either wiped from my memory or buried as self-preservation.

I sat up with a jolt when the doctor told me she had two other patients undergoing regression therapy for alien abductions. She couldn't tell me more due to confidentiality, but she did mention they were locals.

Locals. I'd never look at a neighbor the same.

She asked me to close my eyes and concentrate on breathing. I did as asked, trying to focus on the air flowing through my lungs. After a few minutes, she asked me to think back to a particular abduction event.

I don't know why, but the night of the snowstorm came to mind. This was the night the young couple from the hotel room next to mine were beside my bed when the non-human creature took me through the floor. With the doctor's calm voice leading me through the memory, I spoke of my fear in the moment, how the creature's arms hurt my ribcage, and my shock at falling face-first into the floor. I confessed every detail I could summon.

"That's all," I finally said, my voice shaking. "I don't remember hitting the floor, but I also don't recall anything after falling from the bed."

The doctor tried an assortment of approaches. I followed her instructions, but failed. I couldn't relax and open my mind. "I'm afraid I cannot help you," she said. "Not everyone can be regressed. Maybe you don't want to know."

The experience was deflating. After all, I wanted to know what happened when we hit the floor. I wanted to see what I'd missed or forgotten. I wanted to know where we went and what happened when we arrived. But still, after all this time, I had no idea.

Is it possible I didn't want to know? I suppose. Part of me felt like an imposter, as if I could never live up to the life I'd lived. As strange as it sounds, betrayal crept through my mind, as if I wasn't meant to share my experiences with the outside world. Would aliens approve of my confessions? I left the doctor's office and never spoke a word of the appointment to anyone, not even Lisa.

If my memories were controlled by aliens, they'd won this round. I was a cog in this machine of conspiracy and silence.

How could I change my fate?

Within that same year, I reached out to several US and Canadian-based organizations that supported experiencers in different ways. I contacted MUFON, ICAR, and a few others. I filled out their online reports and submitted my testimony, but never received a response. I emailed Allison Coe, a past-life regression therapist and respected

Quantum Healing Hypnosis Technique Practitioner, in hopes she'd consider me as a client. I'd researched her background and felt that her connection to Dolores Cannon, a woman who had devoted her life to exploring the unknown, might help me.

Every step I took felt like a step backward. I'd spend weeks on highs, rushing to my email inbox in hopes of a response, only to be let down. Perhaps I wasn't very good at relaying my message or sounding believable. Maybe my communication skills were lacking, or my situation wasn't compelling enough. I tend to sound very matter-of-fact. I'm not one to embellish. I am, after all, an electrician, not a writer. And I imagine individuals and organizations like these get a lot of jokers.

Eventually, I gave up. Every abductee gets through the experiences in his or her own way. Some seek solitude in relaxation exercises or meditation. Some find a deeper connection with their faith or prayer. Some find counseling or medication is the answer. Most, I suspect, are best served by finding someone supportive—even if that person doesn't truly understand. Speaking freely, while difficult in many ways, can be liberating.

For me, my family has been that raft. While Lisa can't relate to my experiences, she listens when I'm willing to share. She never judges or ridicules me. My kids were adults in their twenties when I confessed all. They had questions yet never doubted my sincerity. They understood why Lisa and I kept my reality quiet—to protect their innocence. My ability to speak openly has brought us closer. Documenting my experiences and writing this book has also been therapeutic.

If you've seen what I've seen, know you are not alone.

While there may never be a perfect time to search for help, know that others out there understand what you've gone through. Thanks to the Internet, more and more credible information is coming forward. As pilots, scientists, doctors, lawyers, military personnel, political figures, and celebrities speak of alien phenomena, putting their careers at risk and their reputations on the line, the path is paved for more to follow. Walls of discrimination can and will be torn down.

I know it. Times are changing. I was eighteen when aliens abducted Travis Walton. I don't recall hearing about his abduction until I was in my fifties and searching for others who might understand my plight. I'll never forget his interview with Joe Rogan. He'd been abducted in front of seven witnesses, all seeing the spaceship take him, yet the police and the public called them liars. Aliens kept Travis for six days before returning him unbroken and in shock.

But what really got my attention was Travis's description of his experience inside the spaceship. How he felt drugged but kept fighting. He refused to let the aliens touch him unchallenged. He was a strapping twenty-two-year-old logger. I leaped from my chair when he described the glass rod he grabbed from an alien to protect himself. A glass rod. Of all the things to describe; of all the weapons one could potentially find in the presence of humans or aliens, he described a glass rod.

After a few minutes of this information sinking in, I was struck by the realization I hadn't fought as Travis did. I never overpowered my captors. I never used the glass rod to protect my body. I don't think it ever occurred to me to find a weapon of any kind. I didn't rage. I never defended myself.

When the moment of shame passed, one truth emerged from the fog. I had been seven at the time.

27

BEYOND WORDS

I'm not sure what I was shown next can be put into words. Perhaps in the hands of a cinematic artist or a talented sci-fi writer my experiences could be brought to life on the scale they demand but, again, I am an electrician and this is not a work of fiction.

On the night of August 22, 2018, I awoke . . . somewhere. I was standing at the side of a white farmhouse. It didn't feel like home. It felt like a prop. But I knew Lisa was in the kitchen doing dishes. I could hear the water running as she rinsed. I'd just stepped out the side door and onto the yard. It was a screen door—the solid door propped open inside—and I remember looking at the side of the house, noticing how the white-painted barnboard ran vertically. I could smell the grass. A breeze ruffled my hair.

A young man was sitting on the green grass to my right. I didn't know who he was, but that didn't seem to matter. He told me to look up.

The sky was bright. It was a cloudless summer day. In the distance, something glimmered. For a moment I thought it pretty; until the sky was on fire, bright orange and red, the pulsating flames growing larger and closer by the second.

I stopped breathing. Was this Ascension?

Disaster spread across the sky as if someone was unrolling a carpet of fire toward us. It seemed to hover—no part of it falling to earth—like special effects gone awry. There was no sound. The entire world had been muted. The air became heavy, dark, the molten sky almost upon us.

Was this the end? I was about to run, find Lisa, tell her not to be afraid, think thoughts of love, kindness, and caring, so she could ascend to a higher vibrational level, another dimension where those with light and good in their souls will survive. I knew this to my core. I felt it. But before I could move, before the tsunami of fire hit us, the vision ended.

I was back in my bed, shaken but alive.

What was I to think? Were they showing me the future? Had I seen how earth as we know it would end? Was this wave of fire the fate of our people, our planet? Was this an uncontrollable destiny, an inevitable fact? If so, what were we to do with this information? How do we process something so destitute, so demoralizing, so catastrophic? Short of paralysis, what reaction could mankind possibly have? What response was I supposed to have?

Or maybe this wasn't a sure thing. Perhaps this was a warning, a sign of the possible or probable outcome should we remain on our current path as a species, poised to destroy our planet. Maybe it wasn't too late to reverse or alter this self-inflicted outcome.

It could just have easily been a warning of something outside our actions, a threat beyond our comprehension. For decades I had felt the extraterrestrial beings I encountered wanted to help me, guide me—I still believe this—but what if they or others in existence have different intentions, apocalyptic plans to wipe humankind from the face of the earth? What if there is good and bad in all species? What if there are wars beyond our realm?

The possibilities were endless. Science shows our planet was hit with catastrophes in ages past. I knew that other civilizations—on earth or elsewhere—have suffered terrible fates. Who are we to think it couldn't happen again, at any given moment? We have no control over the sun that dictates our planet's orbit and provides warmth and light to everything we hold dear. What of galactic explosions, meteorites, and any number of cosmic hazards we know of—not to mention those we can't comprehend—hitting earth? Is it so far-fetched to consider natural disasters that annihilate everything in their wake? Should I assume this vision shows us what's happened before and will happen again, so we need to prepare? If so, how do billions of people spread across a vast planet prepare for such an event? Is survival even possible?

I was overwhelmed by the implications. Yet, as I lay in bed that night, I couldn't shake the feeling I shouldn't be afraid. Maybe this wasn't an event for everyone, mankind. Maybe this was something only I and others who share my history will witness—others ready for Ascension. What if my human mind jumped to conclusions when the vision was merely a harmless visual to show me how vibrational changes occur? I wasn't shown actual harm done to the planet or its inhabitants during or after this blanket of molten fire—only that it happens. Like most, I'd seen my share of the-world-is-ending movies. Was I jumping to earth-shattering conclusions based on those experiences?

What I do know is that I've been shown things, places, life, and worlds beyond our own. I am not qualified to say whether these are existing dimensions or realities, other planets, or the future. Yet the message is clear: there is more.

Our planet might be slated for disaster. Change could be coming to our species and those who exist among us. The end could already be written, happening one Sunday in August.

Or, Ascension might be personal, exclusive, for me and those ready to move onto a higher vibrational level.

So, where do we go from here? What do we do?

What do I do?

28

THE MESSAGE

It was mid-September 2018, and I was still reeling from a year of life-altering experiences. Seldom did a day pass that I didn't dwell on the changes to come. August will forever be a month I walk on eggshells. The line between excitement and fear was thin. And at sixty-one, I wasn't getting any younger.

I was sitting in bed listening to music one night. I remember being fidgety, waiting for Lisa to finish work downstairs and come to bed. I had my eyes closed, my head back, resting against the headboard while the music calmed me.

A being appeared in my mind's eye, floating in front of me. On impulse, I opened my eyes and sat up.

The being was gone. There was no one in the room but me.

I took a deep breath and settled back into the headboard. I turned up the music and closed my eyes.

It was there again, floating in front of me, an alien that appeared to be a human female. I reached out to touch her, but my hand passed right through. She wasn't solid or formed.

"Be not afraid of my words," she said, "for you asked for answers to your new awareness." Her voice was clear, almost elegant. "I bring you clarity. Clarity is not to be feared but experienced."

She had human features like hair, two eyes, and a mouth. Her nose was arched as ours, only longer, much longer, and when she moved in a way I could see her profile, the fin protruding from the top of her head was pronounced.

"Words do not carry that vibration that experience does," she said. "You need not worry about why; you need only to experience the now."

I listened to her words with laser focus. I needed to absorb everything she could tell me. I had to remember it, write it, and document it. She was explaining the mystery of their ways.

"Many will continue to experience life after life and gain new awareness. They have much to feel, much to learn. Like you, their time for clarity will come, and then they will move to this higher vibration of consciousness. For each, their time will come."

Relief flooded over me. Whatever was in store for the planet, life would continue—humankind would have a place in the universe.

"We, too, have climbed this ladder of consciousness," she said. "Therefore, I tell you this: the level of understanding you seek is not possible now."

She told me that our world is an expression of collective worlds. The expression reflects our state of awareness. Every being in our world is on this path of Ascension, but we are at different stages. We are not judged by our actions, but by our hearts. "It is your heart that made us aware of you," she said.

She paused for a moment, considering me before continuing. "This and only this put you on your path of Ascension. You have walked through time and your scars have been many, but your heart is glowing with compassion, love, and caring. You have found the path of forgiveness and allowed your heart to fill with the light of love."

A warmth spread through me. Her voice, her message, seemed to flow through my veins.

"When your time here ends," she said, "a new expression of life will begin as it always has and always will. That is the path that humanity has traveled for as far back as time can see. You need not look outward for your answers. Your truth is within your heart."

She disappeared and my vision went black. As quickly as she had come, she was gone.

I opened my eyes and crumpled on the bed.

It was a lot to take in. But I had to remember. I scrambled for the notebook and pen I kept beside my bed. I flipped to a blank page and recorded her message. Most of her words were crystal clear in my head, as if planted there, rooted for life. As I wrote, I felt more and more connected to the universe than I'd ever thought possible. I was euphoric.

I could see my life spread before me, how I'd grown and changed from a shy young boy to a scrappy teenager, and how, now, I was a mature man in a whole new skin. I'd become far more caring, loving, and forgiving over the years. I'd felt things I never thought I'd feel. I'd learned empathy for the less fortunate. I shared the emotions of other people, feelings of extreme sadness, loss, or hopelessness, as though they were my pains to carry. Violence hit me hard, internally, like a wound. The older I got, the less I could tolerate television shows or movies containing violence or killing. I struggled watching the news. The thought of someone or something in pain, either physically or mentally, was too much to carry. It physically hurt. I felt compassion on a whole new level.

I'd raised my children with open arms and a ton of laughter. I tried my best to be there for them whenever they needed me. I gave my heart and soul to my wife. My family helps me remember what I've accomplished in this life. I'm proud of our bonds.

I've learned to care for others and to love unconditionally.

There was a time I felt the abductions were horrific experiences that earned me pity I never received. *Poor me, what an awful life to live.* I was so focused on why they had chosen me, I hadn't thought to ask, why not me? Now I know I was granted a gift.

We are here to learn. Our world is a classroom where we experience the wonders of life. Earth is where we learn to taste and smell the riches around us, touch our loved ones, hear and witness the beauty of nature, and feel how emotions are immensely powerful. It is where our souls become better human beings.

Welcome it. Embrace it. Ascend when you are ready.

You'll know when the time comes. Whether you have teachers from another walk of life, aliens who take you under the cloak of night, or you find your soul on your own, know it's there, waiting for you. It only needs nourishing.

29

AQUATIC BEINGS

Only two nights after the woman with the fin protruding from the top of her head appeared to me with her message, I believe I was shown where she lived.

I found myself standing beside a cottage. The all-white cottage was positioned high atop a sandhill overlooking what I assumed was an ocean. Surrounding the house was a four-foot stone wall with a wood gate. The sky beyond the wall was clear and blue. The sun was hot on my face. My toes wriggled in warm, soft sand.

I could hear laughter and chatter coming from below, so I made my way to the gate and looked down toward the water. The sight took my breath away. The sand on the shoreline was completely white and the water was bright blue. There were hundreds of men, women, and children frolicking in the water and on the beach. Their exuberance drew me in.

I followed the steps leading down to the beach. The stairs were wood, the boards flexing and creaking as I stepped. The wood and sand rubbed beneath my bare feet. I think I was wearing shorts and a t-shirt, although I can't be sure.

The breeze was cooler by the water's edge, and when I got to the bottom of the stairs, I turned to look at the white cottage atop the hill. From this perspective, I could see more cottages dotting the landscape for miles. They were built among the sandy hills in various shapes and sizes, similar to cottages I'd seen before. Their muted, neutral colors seemed to meld with their surroundings.

As I walked toward the people playing on the beach, I noticed something peculiar. Their faces were slightly narrowed, their heads more oval than round, and something like a bone or fin stuck out from their heads.

I was in awe. While they had the body parts of humans—two arms, two legs, a torso, and a head—they were obviously another species. The men were slightly larger in height, their build somewhat muscular, but their chests were flat like a human boy's undeveloped chest. Their chin and forehead protruded outward, making their heads seem big, and the fin atop their head was positioned low and to the back. I immediately recognized the profile. They looked exactly like the creature I had seen at the end of the fanfare—the creature who told me he loved me.

This was the same species.

The females looked more like human women in shape and size. Most were slim with elongated, narrow heads and long, arched noses connected to the bone or fin protruding from the top of their heads, close to their foreheads. Their eyes were larger than ours, larger than the males of their kind, and darker.

There were kids of all ages in what looked like different stages of development. They looked the same as the adults, only smaller, and the gender differences were obvious.

They walked like us, moved like us. Even their clothing was human-like and fashionable. They didn't wear bathing suits, but their outfits reminded me of the cover-ups human women wear by pools.

I couldn't tell if these beings could see me or not. If they did, no one paid me any mind. They were enjoying the beautiful day.

Two teenagers swimming in the water rose to the surface only forty feet or so in front of me. The girl was trying to touch the boy, but he kept moving away from her, laughing as she tried to grab him. They jumped and played in the water until the boy dove below the surface and the girl followed.

I stood there watching, reminded of my kids and how they joke around and play when swimming, but after a few minutes, I started to panic. They hadn't resurfaced. I searched the crowd, frantic to find someone concerned, but no one seemed to care. *What if they were in danger?* Without thinking, I waded into the water. It was cold! Not like a chilly, early summer dip, but a heart-stopping icy plunge.

I have no idea what possessed me to keep going, but I sucked in a huge breath and jumped into the water, the extreme cold biting my skin and stealing my breath.

I swam to where I'd last seen the teenagers, then treaded water to stay afloat. This ocean was crystal clear and deep, very deep, even only fifty feet off shore. My arms and legs were tiring quickly. My teeth clattered in my head. I worried I might die in that water—that hypothermia was a very real possibility. But something kept me there, urging me to keep looking. I took another breath and ducked under the water.

There was a whole world under there!

The finned people were swimming underwater like seals. They darted every which way, chasing each other, playing, their clothes clinging to their bodies like down or fur. Although the water was clear, it distorted my vision, and I couldn't tell how far down they were. There was no way I could swim to them, even if the water was warm. It seemed like some were playing on the sandy bottom of the ocean. They were having a blast. And then it hit me: these beings could breathe underwater.

I was stunned. It took several deep-breath dives for me to understand what I was seeing. But the cold was eating at me. I had to head back to the beach.

By the time I pulled myself out of the water, I was happy to be alive. I made my way back to the stairs that led to the cottage and collapsed on the bottom stair, relieved to feel the sun on my skin. I kept expecting these creatures to approach me, to react to my presence, but they didn't. They didn't seem to care that I looked nothing like them or that I was there at all.

What was this place? Was this another dimension on earth? Was this another planet that looked like earth?

I remember sitting on the bottom step, feeling like I didn't belong, like an intruder watching a special moment between loved ones. I was the creature, the outsider, the extraterrestrial.

Was I lost again? There were no holes in the sky, no visible ways to slip from this place. How would I get home? Why was I here? Who or what was showing me this? Did these creatures simply want to show me their home, where they not only exist but thrive?

I sat there and watched for quite some time. So many thoughts were vying for attention in my head.

While this aquatic visitation took place in September 2018, and I've had a handful of extraterrestrial encounters since, there has been a decrease in the number of times they've come to me. The abductions are slowing, winding down as if they've shown me all they've got. Maybe they are no longer interested in a human form reaching expiration. Perhaps they think I deserve a break.

Or maybe, just maybe, they've taught me all they can, and I need to take it from here.

30

CONSCIOUS PUSHBACK

On February 24, 2020, Lisa and I were in Las Vegas. We were on a seven-day vacation, the first three days of which were at Pebble Beach. Lisa was a keynote speaker at a conference there. She was talking about the legalization of marijuana and its effects on the insurance industry in Canada. Despite my struggle to shake a major lung infection, I was thrilled to watch her in her element. After the conference, we rented a car and headed for the desert.

While it would be cooler to say we were in Vegas to gamble, drink, and party, we weren't. Those aren't vices that entertain us for more than a few minutes. Lisa had been to Vegas before. I hadn't. We just figured the downtime would do us good—me especially. Breathing was arduous, and my limbs felt like sandbags. I was on my third round of antibiotics for pneumonia and was still coughing up green phlegm.

The drive was long and hot, but I'd slept a portion of the way and felt a slight jolt of energy upon arriving. The hotel was impressive. We checked in at The Venetian Resort, then hit the town for dinner. I was in awe of all the lights, the sheer size of the resorts, and the energy on the sidewalks. No one suspected we were on the cusp of a pandemic then brewing across the ocean—us included. It was a Monday night and the city was hopping.

Did I mention we're not partiers? We were in bed by ten. Lisa, thoroughly exhausted, was sleeping beside me. I was having trouble nodding off. The lights were off, the curtains drawn, but the safety lights on the stairs leading to the lower lounge off the open-concept bedroom area shed just the right amount of light to keep my attention.

I must have drifted to sleep because I was jolted awake by a vibration that hit strong and hard and without warning. When I shot up in bed, the being was on the left side of me, between the bed and the wall. He was leaning toward me, somewhat over the bed, and the shock of him—what I saw—made me react in a way I had never done before. On instinct, I fought back. With my mind.

I don't know how else to explain it, but I pushed back mentally. It was a big push, a powerful push, a cry, or a warning. *Get back*, I'd screamed without uttering a word.

The being jumped back, away from me, as if I'd decked him. I literally felt him recoil.

He was afraid of me. His emotions filled the space between us like something you could touch. This creature was scared I would hurt him. I felt awful.

"I am so sorry," I whispered. I actually sat up and apologized profusely. "Did I hurt you? I didn't mean to. Please, please accept my apology."

His fear seemed to dissipate, shifting to confusion. He moved closer to the bed. He understood, I think. I tried to radiate sincerity. He leaned over me in apparent curiosity, now sending vibes of kindness, and I attempted to reciprocate with my mind.

This being was spectacular. He was made of what looked like thick chrome wire. There was no start or end to this wire, which was as thick as my wrist, and it seemed to flow, to mold with his shape as he moved.

He was short, only a few feet above the bed. All I could see was his shape: a head, shoulders, and chest, not so different from ours. Nothing but darkness was between the wires—no bones, organs, or muscle. He didn't have arms that I could see, and I didn't look to see if he had legs; I was too focused on his eyes.

His eyes were the size of pennies, like tiny balls of orange-yellow light. They seemed electronic in some way, like a button on equipment, only floating without a connection. There was nothing human about this thing's eyes, and he didn't have lips, a nose, or any other facial features, yet his emotions were real. Very much alive.

"You are beautiful," I said.

I am, after all, an electrician, and wires are my thing.

He didn't smile or respond in any physical way, but I felt his vibrations of friendship and unity. The emotions, although familiar, were much stronger than anything I'd felt with friends or family. This was human emotion on overdrive.

Although my eyes told me he wasn't a living thing, he moved and acted as if he was. Maybe he was a robot or machine with conscious abilities. I considered it a projection or vision. My gut said no. He felt real. He seemed to look me over, then looked me in the eyes. He said nothing and I was tongue-tied.

Seconds passed in silence.

Then, he backed away and disappeared into the wall as if he'd never been there. I leaned forward onto my knees and lowered my head to my hands. The covers were pulled taut from Lisa's still sleeping form beside me. I wanted to wake her up, to tell her all about it. But I couldn't, for all the usual reasons.

I sat there trying to make sense of the encounter, as I always did. The more I thought about it, the more I realized I had learned something. Somehow, I'd managed to create a force, as if I'd pushed an energy shield with my mind. I had no idea I could do something like that. I had never done it before. I'd never physically affected a creature with the power of my personal vibration. It was like harnessing a superpower.

And yet, even at that moment, I knew it was something related to my consciousness, something spiritual in nature. Memories flooded me, thoughts of encounters with extraterrestrials of all shapes and sizes. How many had presented skills humans consider paranormal or magic? All or most could communicate through the mind. Telepathy. Several could move objects like medical tools without physical contact. Telekinesis. Most shared visions, images, and events of the past, present, and possible future. Clairvoyance. All had the ability to calm or control me without physical restraint. Mind control. Over the years, I'd seen alien beings appear from or through inanimate objects like doors, walls, bedsheets, and furniture. Manifestation. I'd witnessed creatures mold into form as they stepped out of my body. On more than one occasion, I'd watched aliens climb over my bed in animal form—some recognizable as animals on earth and some indescribable. Shapeshifting. They didn't move or communicate like any animal we know, but they presented an intelligence I could feel and accept.

This visitation prompted me to consider our conscious abilities from an entirely new perspective. I'd often been in awe of my alien captors, but I'd seldom considered my own abilities. I always assumed they, creatures from afar, were showing me something, teaching me visually. Maybe they were, but not in the way I'd imagined. Was it possible abductions involved the passing of information to alter my state of consciousness? What if they were showing me what I was capable of on my own?

I had, after all, left my body at the age of seven when I'd drowned in my aunt's pool. Levitating above the pool, I existed in a spiritual or energy form. I saw no light or dead family members, yet security and peace engulfed me. This was a near-death experience, yet many people— but for those who have experienced such a thing—believe an afterlife is impossible. As a teenager, I floated away from my sleeping body to roam outside. I made the choice to project outside my physical form, even though I didn't consciously know I could do this. I went through walls. Frightening my girlfriend and her sister wasn't a great call, but my mind was sharp and present. I was capable of astral projection. As an adult, I'd tinkered with my ability to use my mind outside the norm. It didn't take long to discover I could visualize objects I couldn't see and jot them on paper. While any talent I had was by no means consistent, I had proven I was apt with remote viewing. I'd miraculously appeared in strangers' apartments, shops I'd never been to, and places I'd never seen. Could this be an apparition? I had even been in more than one place at the same time, physically and mentally, which is bilocation.

And I am not alone. It took me the better part of five decades to learn that abductees around the world report of thought travel, floating in, out, or around their room and home. They talk of being transported through solid surfaces like walls, floors, doors, and furniture as if solid form is a manmade construct. Some claim to be invisible during an abduction or present in two places at the same time. Many abductees sense or believe their experiences occur in another space or time, another dimension, or that time is not as we understand it. They claim aliens are capable of shapeshifting, often appearing as animals or objects. And words are spoken without the use of a mouth. Thoughts and emotions are shared in whatever language the abductee understands through mind-to-mind communication.

Are these superpowers so far-fetched? Government agencies across the globe don't think so. The US government, for example, uses remote viewers, clairvoyants, psychics, telepathics, and other spiritually talented professionals whose extra-sensory perception works above normal. Military and intelligence agencies recruit these individuals to assist with investigations and strategic planning. While it's not clear if these people volunteer to help governments or if they are coerced and forced, civilians with paranormal talents are evidently considered intelligence assets of value.

Still, something felt wrong, like I was missing the big picture. A nagging feeling suggested there was more to it—much more than my limited interpretation of psychic powers. What was all this for? Why would these abilities exist at all, when society didn't allow me or anyone to use them? For what purpose would a person need these unique skills?

I considered my gifts on a simpler level. While it felt immensely awkward to think about myself in this way, I couldn't deny who I was. I have always been a giver. I have always been led by love and compassion. I've always had a gift for tuning into the feelings of those around me. From a very young age, I could feel people's emotions. If someone was sad, I could sense this immediately, and my natural reaction was to offer some form of kindness. If people around me were happy, I carried that joy to others. An abundance of empathy often led me to help people—even strangers—out of the blue. I've always been calm, patient, and friendly.

I was usually teased for these talents. Even my wife and kids found my need to care for others frustrating at times. I'm sure that watching a grown man cry or refuse to watch anything with violence, destruction, or killing could be comical. Yet now I realize these are gifts I should appreciate. Even when they didn't seem to help.

In late 2019, I was stationed at the Pickering Power Plant. My crew and I had been working together on and off for years, and we knew each other well. For months I'd felt this strange pull toward a guy I'd known a long time. He was younger than me by a good decade, but we clicked and gravitated to each other when out for dinner with a group of work friends or with his little boy. He was easygoing and friendly, a generally happy guy, and we enjoyed each other's company.

Lately, however, something was off. While he seemed fine on the surface, his sadness tugged at me in a way I can only describe as a subconscious draw. I could feel it, internally, like there was a rope connecting us. His hidden sorrow was sometimes overwhelming, forcing me to look away for fear he'd see the emotions on my face. I figured he needed a positive outlet, a chance to shed his heavies, so I shook the feeling and smiled. I didn't want to add to his despair.

Months passed, and this feeling only grew. I didn't know what to do. When I tried to talk to him about anything deeper than day-to-day chitchat, he clammed up. When I got too close, I think he sensed our connection and retreated. I knew to my core he was struggling and needed help. I reached out via text, but he didn't respond.

I knew he was considering suicide.

Out of desperation, I finally went to our boss. "Please do something," I begged. I tried my best to explain how I was aware of this coworker's state of mind, but I'm sure I came across as foolish.

"He's fine," he said, "you're reading something out of nothing." He reminded me we were a tight crew, and everyone would know if there was reason to worry. He sent me on my way.

Weeks later, my boss and I were waiting at a coffee shop when I blurted out my concerns for a second time. "He's going to hurt himself," I pleaded.

My boss told me I was crazy. Our friend would never take his life. "I would know," he said, "our kids play hockey together. Our wives socialize. You're worried about nothing."

A few days later, the frontline manager pulled the crew into the electrical shop for a talk. Before he even uttered a word, I knew what he would say. I'd seen it in my mind. I'd felt it in my bones. I had to hold my breath to stop the tears from welling in my eyes.

Our friend had hanged himself.

Even now, two years later, my heart throbs to think of him and his family.

Perhaps humankind has always had these abilities. Maybe we were once capable of using our spiritual consciousness and have let technology replace our innate skills. There's evidence of this if you consider ancient mythology and folklore of lost civilizations. Stories of Atlanteans and their powers are still told today. For those who study tribal rituals and the generational sharing of our Indigenous People, the human connection to mother earth and everything living has not been forgotten.

Research has led me to understand that many abductees claim extraterrestrials have expressed concerns the human race's spiritual consciousness is out of sync with technological development. They've been shown visions of earth wasted by pollution and toxic clouds or destroyed by nuclear war. They've witnessed humanity through alien eyes, how we kill our fellow man and abuse our planet's natural beauty. Some are told of alien planets destroyed or abandoned and lessons learned from afar.

While I was not shown these things—not that I remember—is it a stretch to think this spiritual consciousness and technological imbalance could lead to the destruction of our species? I don't think so.

It is common for experiencers to believe aliens are trying to change humanity's worldviews, our destructive behaviors, and how we fit within the cosmic circle of life. Most abductees say aliens do not want to coerce us, only change our consciousness, so we choose a different path. What if this is true? Abductees have been told that humans, along with all lifeforms, possess a collective consciousness. They believe in choice and the power to create change. Is it unrealistic to consider that aliens are altering our consciousness to reduce our aggressive nature and stop our mindless destructiveness?

Considering that most abductees talk of learning or receiving information from their alien captors, perhaps the experiences are less about physical attributes and more about spiritual transformation. Many experiencers feel they are participating in a life-changing process of profound personal growth. Most demonstrate a commitment to make conscious decisions from the heart, from a place of love, and to forego negative impulses like hate and aggression. They often feel a oneness with creation, a universal connection with all beings, and expand their view to include entities beyond our planet's borders. Sounds like a good plan to me.

Could it be that simple? What could we be if we nurtured our superpowers like love, kindness, and empathy, employing these positive emotions in our day-to-day lives? What if we tried to raise our vibrational level and connect with others in ways that we find uncomfortable? What if we blew the roof off our stigmas and social limitations and became actively part of *the more*?

As John E. Mack, physician and Pulitzer Prize winner of *Abduction*, so eloquently put it, "The empirical methods of Western science rely primarily on the physical senses and rational intellect for gaining knowledge, and downplay feeling and intuition, and were developed in part to avoid the subjectivity, contamination, and sheer messiness of human emotion. Yet, the cost of this restricted way of knowing may be that we now only learn about the physical world with only limited use of our faculties. In order to learn about the worlds beyond the veil, as abductees put it, we may need a different kind of consciousness."

I'm game. What will it take for you to feel the same?

What if the elevation of our spiritual consciousness is vital to humanity's survival? What if our planet has value or importance beyond our mere existence, and our decisions as human beings have ramifications for other cosmic lifeforms? In simple terms, what if beings from other planets need homo sapiens to get their shit in order?

Recently, I stumbled upon a quote by Whitley Strieber, a fellow author and abductee. "Looking back at my experiences," he said, "I cannot say I felt inferior to them. On the contrary. The creatures I encountered were wise but simple. They not only feared me, they seemed in awe of me. They even asked me what they could do to help me stop screaming. They are not all-powerful superbeings. They are frail, limited individuals far from home."

For the most part, experiencers consider their alien counterparts powerful in inhuman ways. Most extraterrestrials don't have our brawn or sheer size, but they have powers of the mind. We do as well—only we've let our thirst for technology and innovation devour our natural talents.

What if we are the ones capable of extraordinary things?

31

NOT MY TIME

Three nights after meeting the chrome creature, I woke to something that saved my life.

It was our last night in Las Vegas. We were set to check out the following morning—not a moment too soon. My body wasn't accepting this third round of steroids and antibiotics. The pneumonia had gotten worse. I was weak and wanted to be home on Canadian soil. I was sure my doctor would send me straight to the hospital for another chest x-ray. There were rumblings of a sickness spreading from China, but the information was new and vague. Still, we were worried about my lungs. Our flight was scheduled to leave on time. We turned in late that night, around midnight.

I woke with a box over my head.

This isn't figurative. When I opened my eyes, my head was in a black box.

My heart raced for a moment, but not for long. I felt sedated, drugged, and I was struggling to keep my eyes open, never mind calling order to my confusion. Was I sitting up or lying down? I couldn't tell. I had no idea if I was still in bed at The Venetian with Lisa beside me or not. I hadn't seen a blue-green light or felt a vibration. Other than the heavy feeling in my chest, the lung infection, I couldn't feel my body and could hardly move more than my eyes.

The box around my head was painted black. I could see where the walls met the top and bottom, the corners maybe twelve or thirteen inches from my face. The box was a rectangle. Lengthwise, the box seemed to run forever. I couldn't see an end, only black space. There was a light source above me somewhere, behind my head, and the dull light hovered at the top of the box, fading toward the bottom. It reminded me of the light you'd get from a decorative wall sconce. It dawned on me the light source was there for my benefit, so I could see what was happening. There was no one else around—human or otherwise—so what other purpose could there be?

There was a hump in the bottom of the box, about a foot and a half in front of my chin. It was a low, smooth bump, but I couldn't see past it. Normally, I would feel rather claustrophobic in a tight spot like this, but my nerves were calm.

I was fighting to keep my eyes open when something got my attention. Smoke was coming toward me from the dark end of the box. I watched the cloud-like substance, white as cigarette smoke, move slowly over the hump. It didn't spread throughout the box as smoke would. It stayed formed and hit my mouth within seconds.

The smoke was cold. I caught my breath in shock. If it had a scent, I couldn't tell. I don't recall the smell of smoke. I could feel it enter my chest. My throat and lungs burned. It was disturbing but not overly painful, and all I could do was watch the smoke disappear at my lips. The burning increased for a few seconds; then I noticed the smoke was gone.

The black box closed in, and I passed out.

When I came to, I had no idea how much time had passed. I was still in the black box. This time, I could manipulate a few facial features and even wiggle my chin, but my limbs were nowhere to be found. It was the strangest feeling, like I was a bodyless man in a box—some magic show gone bad.

A few minutes later, smoke came over the bump again. This time, I wasn't as nervous. The burning sensation was just as strong but bearable, and I closed my eyes to let sleep take me. I couldn't even hear my raspy breathing.

This process was repeated three times. In that span, I couldn't say if minutes or hours had passed. I couldn't tell you if I was in bed or on the moon. All three times, it was me in a box with smoke.

I woke the next morning, curled up to Lisa: We were still at The Venetian.

There was no sign of any black box or smoke, and I could move my body without issue. In fact, I felt great. I was still quite weak, but my lungs weren't nearly as sore as they'd been for weeks, and I wasn't coughing up anything gross.

By the time we got to the airport, there was a vast improvement in my health. Even Lisa was shocked at the difference.

It was another day or two before I could say the pneumonia had left my lungs, but I stopped taking the doctor-prescribed steroids and antibiotics.

My friends from afar had cured me.

As the days turned into nights and mid-March arrived, my mind was afire with thoughts about the Vegas visitations. COVID-19 was on everyone's tongue. In Canada, businesses were closed, masks were required to enter buildings, and the entire population was ordered to quarantine at home. People couldn't see loved ones who lived outside their household, and if you had to leave home for essential reasons, a six-foot distance was needed between you and others. Airlines were shut down, borders were slammed shut. We were in a full-blown pandemic.

Someone, somewhere, wanted me alive. Perhaps the chrome creature had been sent to check my medical state and didn't like what he saw. Maybe he was meant to take me somewhere or teach me something, but he realized my health needed attention.

I could be making assumptions, but, come on, the timing was too convenient. If aliens didn't cure me of pneumonia to protect me from COVID-19, why the sudden interest in my lungs? I'd been fighting infection for months. There was even a chance I had COVID-19 and not pneumonia. Human doctors weren't testing for a sickness they knew nothing about. It would be months before most medical systems even understood enough about COVID-19 to make a diagnosis and even longer to learn what the virus was capable of.

Coincidence is a possible answer. Or, maybe our planetary friends know more about our world than we give them credit for, and I'm part of a bigger picture. Either way, I am forever indebted to the black box of smoke.

And I'm not the only abductee to think such things. Searching through online abduction forums and credible abduction books has opened my eyes to the positive interactions human beings have had with extraterrestrial lifeforms for decades—perhaps centuries. I've read countless testimonies from experiencers who feel their alien counterparts have helped them. Some feel extraterrestrial medical procedures have saved their lives or improved an existing ailment, and some are convinced their abductors have saved them from life-changing emotional devastation.

Even those considered famous for their abduction experiences have more to tell than the one night or experience blasted through media outlets. If you listen to these people, experiencers like Betty Hill, Travis Walton, and Whitley Streiber—just to name a few—you'll quickly realize their extraterrestrial connections continued throughout their lifetimes. This is the opposite of most media depictions of alien encounters or abductions, which are almost always one-off events threatening humanity. Yet, a large percentage of experiencers believe they've been well-served by beings from other planets.

I'll never forget Travis Walton telling an interviewer he has no doubt the aliens who abducted him were trying to save him. He'd stepped too close to their spacecraft, and some sort of radiation or energy reaction released from the ship hit Travis hard, causing extreme damage to his body. In his mind, the six days he spent on that alien ship were to fix the damage done to his internal organs—possibly saving his life. Sure, he was afraid to learn aliens exist. Of course, his reaction was to fight for his life. But had the aliens left him there, in that field, zapped with who-knows-what, Travis Walton might have died.

This is not the story told by the media. Their story has a much different spin—one meant to appeal to our dark side. The truth is manipulated for entertainment and profit.

Travis's extraterrestrial connection didn't end there. I've heard him talk about aliens taking part in his life ever since. If you truly listen, he explains how aliens changed his life for a second time when he was a grown man married with children.

While soundly sleeping one night, he opened his eyes to find he was running from his bedroom down the hallway. In an all-out sprint, he was mid-motion, unaware of how or why he was out of bed. He only knew he had to get down the hall to his toddler son. He described it as a profound feeling, an urge beyond his control. He knew to his core that his baby boy was dying.

His son was hanging from a structural rail of the top bunk by his neck.

Somehow, while sleeping, the little guy had slipped between the rails of the upper bunk bed he shared with his brother and got his head caught between the bars. The bar was across his throat, his body hanging limp, and he wasn't making a sound. Travis pulled his son from the bed frame without a second to spare.

The boy hadn't struggled. He didn't cry out in the night. Even Travis's other son, who was sleeping below, hadn't seen or heard a thing. Travis's son was dying a silent death until Travis was set in motion by something or someone beyond our understanding. Call this intuition; call it a miracle. But Travis believes, without a doubt, "they" saved his son. As I watched him tell this story, I could see the truth in his expression—his emotions were so visually connected to the event that tears came to his eyes. While the thought of having some sort of implant—a direct line to his brain—or creatures watching his life play out through some mind movie made him fidgety, he was grateful just the same. Aliens had saved his son's life.

If aliens were cold, calculated beings interested in nothing more than our physical or reproductive abilities, why would they help us? If we were merely subject matters in a massive cage, why bother to improve our state in any way? If they didn't have a personal interest in us, a connection of sorts, why make us more than meat on their operating tables?

Why do they help us?

I doubt anyone can answer this unequivocally, but Dolores Cannon studied this subject extensively, and I found her perspective fascinating. She said every living thing has a soul, an internal power source, including lifeforms on or from other planets, and we are all connected by this source. Dolores, and many others who have studied possible links between humankind and alien beings, suggest our souls are shared as energy sources drawn to a place or dimension after physical death, a place sometimes referred to as "home" or "the source." This is a place of reconciliation and contemplation, where souls think about lessons learned while embodied in physical forms—all living forms.

If aliens know that we are all connected, that our souls move from lifeform to lifeform, existing for hundreds, perhaps thousands or more years, could this explain their interest in helping us? What if the extraterrestrials who contributed to my well-being weren't just saving my human life and physical form but helping my soul through pivotal experiences? What if, in turn, my soul's experiences advance the collective, assisting every soul from the source?

There are people who believe the destination of our souls is not random, that our souls choose a path to learn and contribute to the whole. Dolores Cannon suggests it's a mutual agreement made within the source. "We've learned that almost all alien abduction cases are actually mutual agreements made prior to incarnating for the purpose of helping one another. Just as we experience amnesia about who we are and where we are from prior to incarnating on earth, so too do we experience amnesia regarding the contracts and agreements we made with others before coming here."

Alien visitors made it clear it was my heart that got their attention. "It was your heart that made us aware of you," I was told. "This and only this put you on your path of Ascension. You have walked through time, and your scars have been many, but your heart is flowing with compassion, love, and caring. You have found the path of forgiveness and allowed your heart to fill with the light of love."

While I realize this could be interpreted in more than one way, what if this message referred to my spiritual heart, my soul? What if our soul is our spiritual consciousness? And what if alien beings had been drawn to my soul long before my current life, the physical form I have now?

I've read that almost three-quarters of souls on earth today are in the early stages of development. While I can't fathom how scientists can claim such a thing, it's not so out there to think the beginner soul may live a number of lives in a state of confusion and ineffectiveness, influenced by man's less favorable characteristics like selfishness, greed, hate, and the need for power. If our spiritual consciousness is elevated by love and goodness and all that is right with human nature, it's safe to say there is an opposite, a path leading to disaster. Souls already on the path to Ascension can, perhaps, change the outcome of humanity.

So, wouldn't this make us worth saving?

Let me be clear: I have never recalled a previous life. I have never had dreams or memories that made me think I was once someone or something else in another life. I also hold no religious belief that would have me lean one way or the other regarding souls.

That said, I suspect we have souls and our bodies are the containers, the means by which we exist and experience the elements around us. Cultures and religions across the globe have shared stories of souls for generations, maybe since the dawn of man. Heck, a weekend of binge-watching Netflix can tell you a lot about our kind's belief in souls, the afterlife, and the connection to our conscious mind. Even science has chimed in on the matter. To this day, many believe the theories of Dr. Duncan MacDougall, who, in 1907, claimed the human soul weighed twenty-one grams.

My research has led me to believe that past-life recall is a common phenomenon among abductees. Many experiencers talk of reliving or remembering past lives or experiences from previous incarnations. Abductees claim their souls have inhabited earth's creatures in addition to lifeforms from other planets or dimensions. They suggest these experiences teach souls to appreciate the diversity and beauty of all life within the universe, so they can share what they've learned and guide other souls through the process. They believe souls don't fully remember past lives by design. Memories would taint every new existence and influence every new experience. To truly experience life in its raw form—the good, bad, and ugly—one must feel it. Souls incarnate on earth to help humanity raise its vibration in the process of Ascension so we can exist in harmony.

To some, this might sound hokey. But what if it's true? There is no doubt alien beings come to earth to interact with our kind. Would their purpose be anything less than monumental in scope?

While I cannot say my perspective rings true for all abductees, I know, deep down, my captors have helped me in ways I only see as positive. Their presence has enriched my life.

As I write this, it dawns on me that this Las Vegas vacation was only two years ago, and I have not suffered one moment of sickness since. I can even be around cats and dogs and allergens without coughing, something I was previously able to do only with an inhaler or antibiotics. Having struggled with asthma and lung infections for a large portion of my adult life, I'm astonished I made it through the COVID-19 pandemic unscathed. Call this what you will, but I consider myself lucky.

I should have known. It was February, not some Sunday in August. It wasn't my time.

32

APRIL 27, 2020

UNEXPLORED WATERS

On April 27, 2020, the Pentagon officially released three UFO videos taken by US military fighter pilots. The videos were cockpit recordings taken in 2004, 2014, and 2015 by US Navy fighter jets from the *USS Nimitz* and *USS Theodore*, aircraft carriers off the US coastline.

When I saw these videos and the accompanying articles circulating with an enormous amount of buzz, I was blown away. Finally, the US government was acknowledging UFOs. They were calling them UAPs, unidentified ariel phenomena, but same thing.

Imagine my thrill to hear Commander David Fravor talk of the 2004 sighting off the coast of San Diego, California. "Control had been tracking these objects for weeks," he said. Radar operators using state-of-the-art tracking devices had been watching these flying objects come in and out of view over the water yet hadn't filed official reports. "Control didn't know what the objects were," said Commander Fravor, "and they weren't a threat." There were no navy pilots in the air, and the objects weren't showing signs of attack. For the most part, they moved slowly, flying at speeds less than 120 knots or 138 miles per hour, and flew in formations or clusters of five to ten. These objects didn't move like any aircraft known to man. In simple terms, they zoomed back and forth, up and down, defied gravity, and disappeared instantly.

Curiosity won on November 14, 2004, when control operators asked Commander David Fravor and his unit to investigate the objects while out on a routine training drill. The two FA-18F jets were unarmed, as per training protocols. They were about eighty miles from the coast when Fravor, along with his weapons systems officer in the back seat, saw a large section of whitewater bubbling on the surface of the Pacific Ocean. Lieutenant Commander Alex Dietrich and her copilot in the second jet also saw the unusual water disturbance, like a churning of the ocean surface. They agreed to remain above while Fravor got a closer look.

As Fravor's visual became clearer, he noticed a strange, white object hovering above the whitewater. "It was about forty feet long," he said, "a smooth, white oblong object resembling a large Tic Tac breath mint." It had no wings, no visible means of propulsion, no lights or markings, and no exhaust plumes.

The object really got his attention when it moved. "It went from a complete standstill to taking off in a flash," he said. "We don't have the technology to do this." The object flew in all directions, at rapid speeds, falling from 60,000 feet to hover at 50 feet above the ocean surface without a sound. "There was no sonic boom as this thing took off, and it was faster than anything I'd ever seen."

It turned to face him. It moved close, mimicking Fravor's movements, then disappeared. All four of the pilots saw the craft simply vanish into thin air. Control saw the whole thing on radar. "When the Tic Tac disappeared in front of me," said Fravor, "it appeared on radar forty miles away in an instant. "We had no view of it flying away, and there is no way an object could fly that fast. Yet it did. It appeared out of nowhere."

This wasn't the testimony of some random or ordinary guy. David Fravor was a commanding officer who served the US military for twenty-four years, eighteen of those flying for the US Navy. Joe Rogan said it best in his interview with Commander Fravor when he noted that military fighter jets cost somewhere in the range of $70 million each. "They don't typically give those to morons," said Rogan. Humble Commander Fravor had to agree. His crew had the same story—they'd seen everything.

Finally, I thought, *the public has witnesses with extreme credibility.* I soon learned there were many more military men and women who saw these unidentifiable flying objects.

Commander Fravor and his crew returned to the aircraft carrier *USS Nimitz*, and a second unit, led by fighter pilot Chad Underwood, headed out. They also found the Tic Tac object flying at incredible speeds, maneuvering in ways physics could not justify. Underwood followed, locking his jet's radar systems on the Tic Tac. He recorded its movements in infrared. His radar received a signal back, actively jammed by the Tic Tac or an unseen source, which evidently had the technology to do so. The object was actively responding.

Anyone can find this video online. The footage appears to depict what Fravor and his unit identified as a forty-foot-long, white, oblong-shaped object hovering somewhere between 15,000 and 24,000 feet with no notable exhaust or propulsion source and no sonic boom. The object appears white on radar because it's hot, the temperature higher than the air and water surrounding it. It appears black when the pilot switches radar mode for a better view. Again, it darts to the left and disappears.

This wasn't the only sighting. These objects appear to be prevalent off coasts, where the military encounters them regularly. Witnesses speak of dozens of these flying objects in different shapes and sizes. Some are triangular, some round, and some are square. Several pilots reported what looked like a large, clear beachball with a cube floating inside. These objects fly for hours on end, sometimes days, without refueling. Air currents do not affect their flight path. They are solid masses that manipulate gravity without resistance. Several have even come close to hitting military planes, almost knocking them from the sky.

Most are discovered over oceans or lakes, and many appear to come from or disappear into large bodies of water.

Retired Air Force Lt. Colonel Richard French testified before Congress in 2013. As a lead investigator working for the US arm called Project Blue Book in the 1950s, French's job was to discredit and debunk hundreds of UFO accounts, including the alien encounter he himself had witnessed. His testimony included a detailed description of two extraterrestrials performing repairs on two submerged spacecraft off the coast of St. John's, Newfoundland.

"They said, 'We have a UFO report, and we want you to investigate it,' and that was standard for what I was doing," French testified. "They told me there were two UFOs involved and that they were in deep water, after entering the water doing roughly 100 miles an hour."

This wasn't a glimpse or a simple misunderstanding. Having been ordered to the scene by superiors, French joined at least a hundred people standing on the wharf, including several local police officers, looking on in amazement. They'd been watching these creatures move about on two circular objects in the water for hours. "They were doing what looked like repairs to the crafts," French stated before Congress. "Then, both ships flew out of the water, straight up, at speeds impossible for any ship or plane we have today."

There are documented reports of unidentified flying objects dating back generations. There are reports from World War II pilots in several countries. Apparently, Christopher Columbus chronicled his crew's sighting of a craft rising from the ocean, then flying away. I've read that the ancient philosopher Plato spoke of ships that could fly in the air and underwater as if both elements were the same. Hundreds of airline pilots have reported seeing objects they couldn't recognize as man-made.

Governments around the world track UFO sightings offstage, and several independent organizations receive testimony from credible witnesses daily. MUFON, a US-based organization focused on uncovering alien phenomena, has hundreds of reports on unidentified flying objects entering or exiting earth's waters. Many can be read online. There is also an extensive list with links on the website Water UFO, which is also linked to MUFON. There are even published books documenting UFOs seen under, in, or over bodies of water.

The connections are there if you're inclined to find them. In my thirst for knowledge, I noted how UFOs and the earth's water sources share a strong connection. The link is something I've thought a lot about.

We tend to think aliens come from afar, from other planets, far-off galaxies, or solar systems. While this might be true of their origin, there is nothing to suggest extraterrestrials fly their spacecraft into our atmosphere to land on earth. This absence of proof does not mean they don't, but knowing aliens come, and visit often, I'm left to wonder how they move about without large-scale detection.

What if they don't come? What if they are here, have been here a while, maybe a long time, and don't come and go from our atmosphere? What if they are stationed somewhere on earth? What if they are here, under our noses, and have been for ages?

Countless witnesses speak of UFOs going under water without so much as a splash. The flying objects reported by many people, including high-ranking military, seem to bend or manipulate earth's elements like air, land, and water as if those who built them know something—or a lot of somethings—we don't. Is it so unrealistic to consider they might have underwater docking stations? Could this be where humans are taken when abducted? Could this be how extraterrestrials hide from us—if one assumes they hide at all?

Think about this. Aliens have the technology to make their spacecraft appear and disappear in the blink of an eye. They can jam or override our most sophisticated radar and tracking systems and fly undetected. For goodness sake, they pass through our military airspace without human interruption, even when witnessed, and don't care that they attract the attention of our armed forces.

When you're done twisting your noodle around that chunk of info, chew on this for a moment.

Planet earth, as we know it today, is 71 percent water. Of the earth's surface water, approximately 97 percent is ocean, and 3 percent is lake water or ice. Now, in 2022, we have the technology to explore less than 4 percent of the water sources that make up over 71 percent of our environment. Despite its sheer volume and the impact water has on the survival of every organism on this planet, our oceans remain a total mystery. Only a very small percentage of our ocean floor has been mapped, explored, or even seen with human eyes. The rest can't be reached—our equipment isn't advanced enough.

In contrast, we know a crapload about the moon and other planets. We've mapped a huge percentage of Mars, and the surface of the moon has been studied at length.

Now, if you were an alien race from afar and needed to hang low on earth, where would you go?

Several times, when I was in my mid-teens, I was taken to a strange place I couldn't identify. I wasn't sedated when I woke in this place, but the memories are fuzzy, as if remembering wasn't part of the plan.

While I don't recall leaving my bedroom on these occasions, I will never forget waking in this big room with maybe fifteen or twenty human boys and girls, all around the same age as me. I'd been here before, several times, and although I couldn't comprehend how I knew this, since nothing visual sparked a clear memory, I felt sure I'd been here before.

The class was a mix of hair colors and different facial and body features, but we were all pale-skinned teenagers wearing the same orange one-piece jumpsuit with no embellishments. A voice asked us to sit at the strange desks, and we did as we were told. The room contained way more desks than there were kids. Like the others, I was nervous, so we sat side by side instead of spreading out. I suspect no one wanted to think about the empty seats.

There were no windows in this room, but for some reason, I knew without looking there was an exit behind me. The space was more than twice the size of any schoolroom I'd ever seen. The desks consisted of a stool for sitting and a flat piece out front, a spot hardly large enough for your arms to rest while you leaned forward on the seat. We weren't meant to bring supplies or take notes. I don't remember noticing what the desks were made of or what color they were, but I remember they felt cold.

I have no idea how long we sat there, confused and silent, but an older lady entered the room and caught our attention. She wasn't wearing a jumpsuit. She looked like any middle-aged Caucasian woman from home. She spoke to us in a kind, polite tone, in what I believe was English, suggesting we needed a break. A break from what, I wasn't sure, but six kids—me included—got up from the desks and followed her out of the room.

The woman led us through a wide opening at the back of the class, and before I even stepped into the hallway, I knew there was a long metal bench further down that hall. I'd sat on that bench before, maybe more than once, waiting for something or someone. I couldn't see the bench at this moment, just the opening that led to a high step down to an open area. It was like a schoolyard, only barren.

When I hopped from the step, following the group, I sank when my feet hit the ground. I don't know if my feet were bare or if I was wearing shoes, but it took me a moment to realize we were standing in sand. It was regular sand, pale brown, with no distinct texture. I can't remember if it was cold, but I'm sure it wasn't hot. The six of us moved to take in the space.

The building we had come from was just one section of a large square compound that surrounded us. To our immediate right was a dark tower. It was about twenty stories high, like an apartment building, only simpler. Flat windows with no trim lined the two sides of the tower I could see from my spot in the sand. Like the rest of the buildings in the compound, the tower was made of something dark, shiny, and smooth—maybe glass or plastic. There were no apparent seams or joints and no bricks or stones to be seen. I remember staring at the tower for a moment, trying to pluck memories I could feel on the fringe of my mind but couldn't touch. I could almost envision the inside of that tower, but I didn't know if I'd been there before.

In the center, where we stood, was a courtyard of beach sand, maybe seventy or eighty feet perfectly squared. There was an abrupt ledge about twenty feet in front of us, where the sand went down a steep hill. "Be careful," the lady said, pointing to where the sand dropped over the hill. "Don't go near the edge." Of course, we all looked.

In the distance, the sand led to a building on the other side of the courtyard. The building looked the same as the one behind us, the place we came out of, only lower to the ground. It had tall rectangular windows that ran horizontally in two rows, all facing the sandy courtyard. From this distance, I couldn't tell if the windows were glass or just openings, but there were no doors, and nothing inside the building was visible. Above the two rows of windows, the exterior walls rose several floors up, unadorned, like an afterthought.

What caught my eye was the saucer in the sky. It was silver—round from below. I couldn't see any lights or markings recognizable as a craft made by man. It looked like a solid mass just hovering in the air as if parked. I couldn't estimate its size from below, but I think it was bigger than the compound. The tower to my right climbed almost as high as the saucer, and I assumed we had arrived on this spaceship before making our way down the tower.

Beyond the saucer, the sky was grey and solid, as if everything was inside a dome of concrete. Light from an unseen source seemed to coat the inner parameter of the dome, dark in spots, but totally artificial. I remember catching my breath, wondering how we could breathe in such a place.

A girl screamed, and I turned in time to see a blur of orange falling over the crest of the hill. She was halfway down when I made it to the edge of the slope. I watched with the other kids as the girl found her footing and pulled herself upright on an angle. *How was she going to get back up?* I looked around for the lady who brought us here from the classroom, but she was gone. The girl struggled to climb the hill, but it was steep, and the sand kept her from moving forward. She kept sliding back down.

Without thinking, I went over the hill to help the girl. I wasn't worried she would get hurt; I was worried she'd get in trouble for going over the ledge. My heart was pounding as I climbed down the hill sideways, taking one leap at a time. The girl didn't say anything as I took her hand, pulling her from the sand. We climbed up on our hands and knees, following the slope as it eased to the left, where we seemed to get a better grip.

When we got close to the top, the other kids just stood there, watching. No one reached out or offered encouragement. I maneuvered over the edge first, then tugged the girl to the level ground beside me.

Once upright, we shook the sand out of our hair and jumpsuit, trying to hide the evidence. I don't know why I thought we'd be in trouble. I hadn't seen anything bad happen. Yet, the hollow feeling in the pit of my stomach said we were lucky the lady was nowhere in sight.

My memory is muddy from here. I don't recall if the woman returned, if we were ushered back into the classroom, or if the girl and I got in trouble. It's as if time from this point on ceased to exist.

I do know this: a domed compound built on sand could very well exist on our ocean floor, where we would never see it. Alien craft can and do fly in and out of our water with ease. I know I've seen creatures who walk and act like us yet breathe underwater. I've watched them play on a beach and swim long and deep in a freezing cold ocean where humans couldn't survive. Some extraterrestrials have fin-like protrusions on the tops or backs of their heads, along with aquatic features. I don't know if I was shown all this through a lens, in real-time, or through a connection of the mind and soul. But I do know the question is worth asking.

Could aliens be stationed under the dark waves of our planet's oceans?

I'm not qualified to say, but if I had the technology to hide on earth in plain sight, that's where I would go. I'd live close to work, where I could investigate and help the human race in their environment. I'd flourish where the technology of my adversaries can't reach. And I'd relish knowing I could come and go as I please, right past the human race's strongest, most guarded security forces, without anyone believing I was there. I'd be invisible.

I'm reminded of a great quote from Brian Cox, physicist, and professor of particle physics. "You need to be humble when talking about science," he said. "The most valuable thing about science is the realization that we don't know. We need to store all sides, all options, on any one subject, in our minds. The possibilities are only limited by nature, and nature is not absolute; it's always learning and changing."

So are we—if we'd only open our eyes.

33

TIC TACS

On June 25, 2021, just over a year after the Pentagon released and verified the aforementioned UFO videos, the US government declassified a report titled *Preliminary Assessment: Unidentified Aerial Phenomena.*

In summary, the report refers to 144 credible UFO reports made by US military aviators and flight staff between 2004 and 2021 (the majority from 2019, 2020, and 2021) and how the government should consider these objects a national security threat. The sources are high-ranking military men and women from across the country who have seen flying objects of unknown origin—craft capable of maneuvers beyond anything our most talented scientists and inventors can conjure—while the events were registered across multiple military sensors, radar, infrared, electro-optical, and weapon seekers.

The report notes information collection issues like a lack of standardized recording systems, technological limitations of our most sophisticated military equipment, and the stigma that keeps pilots and their crews from reporting sightings. It suggests government funding is required to further study the multiple types of unidentified objects that cluster around military training and testing grounds exhibiting an array of gravity-defying ariel acrobatics and posing flight hazards for military aviators.

This report, released by the Office of the Director of National

Intelligence, is available for public viewing, yet has received limited media coverage. Sure, extraterrestrial phenomena are being shared more via mainstream media and online forums these days, creating world-wide interest, but not to the extent one would expect. You'd think this news would be splashed across the front page of every newspaper and magazine in the world, on the tongue of every person: US Government Admits UFOs Are Real.

Nope. The circulating version of this report is nine pages. Nine pages. Zero mention of alien beings. Who or what, pray tell, do they think is flying these objects?

The biggest concern raised in this report is not that mankind isn't the smartest jelly bean in the jar. It's not that something, somewhere, had the knowledge to build these flying machines in the first place. And it's not that aliens in flying vehicles shoot about our airspace like pinballs in a free-for-all pinball game without a soul demanding the world take notice. The biggest concern raised in this report is that another country, such as China or Russia, could duplicate alien tech-nology first.

You read that right. Of all the wondrous possibilities this intel could uncover, the US government is worried the technology will fall into the hands of humans who will then have a leg up, so to speak. Even the media reports and interviews focus on technology.

"This technology is a game changer," they cry.

"The country to get this technology first wins," they say.

They don't consider what the confirmation of otherworldly machines flying around earth means for humanity. They don't ques-tion what the creatures who fly these machines could teach us, tell us, or show us to better our planet and our people. They don't discuss how creatures from other planets or solar systems might have cures for our sick and dying. These people don't wonder if alien lifeforms could tell us how to get along so we don't kill each other off, bloodying our hands with our fellow man. No one asks if extraterrestrials could help us save our planet from ruin, from being robbed of its resources. Not one person asks why the hell they are here.

No, all they want to know is how to harness alien technology to gain more power over other human beings. They want weapons.

If I need to point out the flaws in this mindset, I'm afraid our kind is doomed.

Look at the evidence. The Pentagon validated videos of UFOs. US National Intelligence reports confirm the government has no idea what these flying objects are or where they come from. Intelligent, honored military men and women have seen UFOs with their own eyes, in addition to domestic pilots across the globe. Hundreds of thousands of civilians speak of UFO and alien accounts spanning decades. All this, yet talk of extraterrestrial phenomena happens in hushed tones by the watercooler, followed by laughter.

When are we going to learn?

I believe John E. Mack said it best in his award-winning book *Abduction*. "We are a species out of harmony with nature, gone berserk with desires at the expense of our planet and other living creatures. The task of reversing this trend is momentous. Even if we recognize the peril we've created, the vested interests that stand in the way of change are formidable. Huge corporate, scientific, educational, government, and military institutions maintain a paralyzing status that is difficult to evade. To a large extent, the scientific and government elite, along with their selected media, control what the human population believes. These monoliths are the principal beneficiaries of the dominant ideology. To governments around the world, the abduction phenomenon presents a problem. It is, after all, the *business* of government to protect its people. Suppose officials acknowledge that strange beings flying radar-canceling craft can defy the laws of gravity and space/time to invade homes and abduct innocent citizens. In that case, they aren't doing their job. This may explain why government policy concerning UFO sightings and abduction reports result in a garbled mixture of denial and cover-ups that fuel conspiracy theories."

So, why now? What happened in 2021 to force the US government to declassify UFO intelligence? While I can only speculate, my money is on surrender. Too many people have discovered the truth. It's harder to bury the testimony of the masses when all of us, even five-year-olds, walk around with cell phones that take videos and pictures in an instant. It's difficult to debase and slander credible sources, especially those holding top government and military positions. Today, too many people are talking about UFOs, even the US Congress, where some are demanding answers. In basic terms, the secret is getting harder to hide.

Countries throughout the world employ officials responsible for tracking and investigating UFO phenomenon. Most operate quietly behind the scenes, without drawing the attention of civilians. Whether this is intentional or not is up for debate. Project Blue Book, the US government's UFO investigation arm, operated out of the Wright-Patterson Air Force base from 1952 to 1969. Various US government branches studied UFO phenomena prior to 1952, but none exist today unless you believe Area 51—the highly classified US military base in the state of Nevada—contains physical proof of UFOs and aliens.

In a 2019 interview, Bob Lazar said he was one of many scientists working in a section of Area 51, recruited to back engineer alien technology. As a physicist, he specialized in nuclear development. "There were at least nine alien ships on site," he said of his time there. "There were ships in different shapes and sizes, and I think they'd been there a long time— for many years. The ships and their reactors were borderline magic, and we don't have the technology to make these."

He talked of the extreme power within these machines and how scientists like him were charged with finding ways to copy the technology. The craft he worked on was fifty-two feet in diameter with a small

interior built for creatures half our size. Three tiny seats faced a center reactor. The sparse space was one cohesive color, metal or ceramic
looking, and melted together without corners or junction points. Lazar
saw them fly on several occasions. He witnessed their radical moves
and gravity control. Apparently, Area 51 scientists could operate some
of these flying vehicles, but they didn't know how they worked and
couldn't replicate the technology.

Maybe that's a good thing.

It's no wonder aliens don't park on the White House lawn to say
hello. They know government leaders would panic, attack, and kill their
kind, and ours, in the process. This is what we do. Yet, I suspect there's
more to their madness. Creatures from afar simply don't operate this
way. I think they prefer to teach us how to fix our problems internally.
They provide us with guidance so we can embrace our inborn abilities.
They abduct us, study us to understand how to help, and gently place
us back into the wild to correct our wicked ways.

They want us to choose to be better people.

Between the Pentagon videos and the declassified UFO report, my
mind was swimming with the potential for change. Although I was disappointed the government and media tiptoed around the big picture,
I still felt the US government stepping forward to validate these claims
was a huge step—a pivotal development for humanity. These sightings
and how governments, media, and the population at large react to them
have the power to alter cultural bias and negate the stigma that wallops
anyone who speaks of flying objects or the creatures who command
them. They have the potential to shift our scientific focus and support
uncensored dialogue, directing mankind to gather knowledge and
understanding for what has been—for far too long—dubbed as insan-
ity, fantasy, or taboo. And, for experiencers like me who have known
UFOs and aliens exist for entire lifetimes, generations even, a door has
been opened.

Memories were pulled up to the surface where I could reflect on moments I'd never forgotten but had buried deep, where no one could use them to harm me.

I recalled the summer afternoon in 1997, around five or six o'clock, when I saw a UFO fly over my house. We lived in Whitby, Ontario, I was about forty at the time, and I was standing on the front lawn with my neighbor. We'd just finished a neighborhood water balloon fight, and the moms had ushered the kids home to dry and eat. It was broad daylight on a sunny day.

My neighbor was a postal worker. His name escapes me now, but we were chatting on my lawn facing the sidewalk. Lisa and I had socialized with his wife Rena on several occasions. She was a stay-at-home mom with a new baby. She'd been playing water balloons with us, but I hadn't talked to her husband much over the years. Our work schedules didn't mesh. We were talking about my brother-in-law, who was also a postman, when he looked up and pointed to a silver object in the sky.

Half a city block away and 100 feet in the air, the UFO, maybe 20 or 30 feet long and 8 or 10 feet wide, moved toward us. It looked like a silver cigar case hovering above the houses. Its surface was completely smooth, slightly reflective, and round at both ends. This object had no wings of any kind. I didn't see landing gear or an exhaust system. There were no lights or windows or markings across its body. It was moving slowly, heading west over our neighborhood, maybe thirty or forty miles an hour, and it made no sound at all.

We just stood in the yard, staring at this thing floating over us.

"What is that?" I said, my voice uneven.

My neighbor shifted his weight from foot to foot. "I don't know," he said.

All my well-honed self-preservation instincts kicked in. "Maybe it's that cruise missile they're testing out west?" Who was I kidding? The government wouldn't be testing missiles over a residential area. And this cigar case was much too large to be a missile.

Is it here for me? Please no, don't take me here, in the middle of the day, in front of my neighbors.

Everything I'd ever worried about came crashing in. For that split second, I saw my life evaporate before my eyes.

Then, the UFO disappeared. One second it was there; the next it was not.

"Where did it go?" my neighbor whispered.

I kept my mouth shut. Nerves had my body in knots. I'm not sure I could move or speak if I wanted to.

"It's . . . gone," my neighbor said, searching the sky. After a second of silence, he looked at me, confusion clouding his features. Our anxiety was palpable yet different. When I didn't comment, he turned and walked away. People didn't talk about this stuff.

They still don't.

I'd never thought of this UFO as a tic tac—not until recently, after I heard the testimony of US military pilots. Now I think it's a good descriptor. Although I hadn't seen a UFO flying in full view before this day in 1997—my perspective was always from the underbelly where the blue-green light made it hard to see—I was struck by our human reaction, the way my neighbor and I responded to seeing something so wrong by society's standards.

This is a perfect example of how people react to alien phenomena. We look, we see, we try to process, and when we realize we can't make sense of it, that we are seeing something we're not supposed to see, something science, law, government, media, and society deems impossible at best, crazy at worst, we dismiss it. We pretend it never happened. *What spacecraft? What flying object? I didn't see anything. You didn't see anything. No one saw anything. It didn't exist.*

Even me, a guy who had been abducted and visited by extraterrestrials

his entire life, followed protocol. I played the part. I reacted as any
human being would in this circumstance.

Why do we do this? Why do I do this?

We need to stop.

I need to stop.

EPILOGUE

My life has been unusual. I know.

And yet, I wouldn't change a single thing. I've seen things beyond logic. I've had experiences most cannot comprehend. I've known fear, rage, pain, sorrow, loneliness, shame, grief, and a myriad of heartbreaking emotions. I've also experienced joy, admiration, acceptance, hope, freedom, and love.

I've lived. I can't imagine a more fantastical gift. Now it's my turn to pay it forward, to share.

For a long time, the working copy of this book was titled *The Truth*. By definition, *the truth* means something factual, proven, or believed. The experiences within these pages are my truth. But in the process of diving deep into my past, into memories I've tried to bury and hide for the better part of a century, I've come to realize the truth is subjective. My truth will be someone else's fiction. It's all in one's history, experiences, and perspective.

While I have no doubt these memories are as true as I can recall them, I'm not blind to the fact that time has a way of altering perception. The things I experienced at a young age are not as easy to see now, through eyes that have endured fifty-odd years of encounters and a lifetime of . . . well, life. In some ways, the passing of time has made memories more vivid. I now understand things I couldn't possibly interpret in the moment. In other ways, my memories are pocked with black holes, details that were either unimportant at the time or merely too much for my brain to cope with.

I can only tell you what I've learned from my time on this planet—most of those years greatly touched by extraterrestrial beings. Here is what I know. Here is my truth.

Aliens exist.

We are not alone in this universe. Not even slightly. There are many different species of what we call alien lifeforms, and several—if not all—live or come to earth on a regular basis. Some look human or part-human. Some are similar to aliens you've seen in the media or movies—the ones based on real-life accounts. Most are beyond our comprehension, indescribable.

Aliens come in different sizes, shapes, and colors. I've never seen a green one, although that doesn't mean they don't exist. They are not always hardy and strong. Most are frail, maybe even fragile. Some are short, less than half the height of an adult human, or even smaller than a human child. Many have torsos, arms, legs, and heads. The ones I've seen have eyes, noses, ears, and mouths—although different than ours and not always positioned where one would expect. Some have hairless, gray skin, and some are covered in fur. Some resemble land animals, and some appear aquatic—fins and all.

I've learned there is no end to the wonders of nature.

Not all extraterrestrials arrive in physical form. The aliens I refer to as extradimensional appear as translucent, luminous, or manifestations of objects and things. They are not solid or tangible, although they might be where they exist. I have no way of knowing if this is their true form or a means of communication.

Aliens are not mere animals—if such a thing should even be said. They have personalities.

They smile, laugh, and banter. Some have a sense of humor.

The beings I've encountered have abilities we humans deem paranormal, superhuman, or downright ridiculous. They use telepathy, mind control, astral projection, remote viewing, and bilocation—just to name a few. These are the abilities I've witnessed, but they could be capable of any number of miraculous feats.

Aliens exist. I've seen them, touched them, heard them, and I suspect I have only been exposed to a very small sampling.

I assume there are more—many more.

They are not here to harm us.

Aliens have abducted and visited me more times than I dare to count. They take me when they want, where they want, and don't offer pleasantries. If they wanted to hurt me, I'd be hurt. If they wanted me dead, I wouldn't be breathing. If they didn't take care to return me to my bed, car, hotel room, or wherever they found me, I'd be . . . who knows where. I could disappear indefinitely and no one would ever find me.

I've been abducted since I was a boy of seven. That's more than a fifty-year span. Not once have they harmed me in an irreparable way. To my knowledge, I have every body part I was born with and no surplus of appendages. Their experiments have left no visible scars, marks, or patterns on my body, and they usually made an effort to keep me calm using sedatives or mind control. I even suspect a large percentage of experiences have been stripped from my memory—probably for good reason—to protect me from things I am unable or ready to understand.

In short, they aren't the bad guys.

Don't get me wrong; we are the prey. They take us without discussion, without permission. Few people, if any, would volunteer. But the experiences aren't always dark and sinister. I presume it's usually the fear of the unknown that creates anxiety in the moment. There is massive conflict between the experience itself and what we expect the experience to be. In basic terms, we are conditioned to be afraid, to fear them, so we do. Even when there are limited reasons to be scared.

Are all extraterrestrials harmless? I don't know. Even the human race has its share of evil. We are not immune to greed, arrogance, deception, hatred, and violence. There are humans among us willing and able to inflict physical and emotional pain without a moment's regret. Is it unrealistic to think these traits aren't universal? Who knows?

I do know this: we are 100 percent defenseless when it comes to threats outside our planet. If aliens wanted to overtake us, they would. If they wanted to hurt or kill us, they could—easily.

Yet they don't.

They are not necessarily smarter or technologically advanced.

I have no benchmark for knowing beings from beyond are technologically more advanced than humans. Our interactions have always been one-sided. I am the subject and they are the leader or investigator. I can only speak for what I've seen. They use transportation devices that look nothing like our cars, ships, or aircraft. The exterior parts of their spacecraft are made of some sort of white or silver-colored metal or material unknown to me. I only know energy, vibrations, and light are involved. I've never seen an engine, propeller, or any component I recognize as a form of movement, yet their crafts fly in ways that defy our understanding of physics. They hover, appear, and disappear in a blink and make little to no sound.

The same can be said for the interior of their spacecraft and the equipment they use during abductions. While I can't be sure the operating or learning rooms I've been taken to are within the walls of their spacecraft, I know I've come to white, mostly wall-less spaces made of light. I've assumed I'm still within earth's atmosphere on an alien spacecraft in these moments, but I could be anywhere—another planet, another dimension, another time. How would I know the difference?

Their machines remind me of hospital equipment, yet there hasn't been a single piece of machinery I could identify with certainty. And I've seen my fair share of machinery—I'm an electrician. I've also never seen an extraterrestrial use human tools or instruments, yet I suspect some understand human anatomy and how to explore us internally without causing obvious damage. Some might even have the medical knowledge to cure or fight human ailments and have obviously advanced on a spiritual or vibrational level, understanding the human mind and consciousness beyond our current means.

Do these things make them smarter or more technologically advanced? I can't be sure. Different, yes, but maybe we're more advanced in other ways.

Maybe that's one of the reasons they find humanity so fascinating.

They are capable of communicating with us, and do.

Some alien races understand our language. I suppose I should clarify. Some have communicated with me using the English language. I have no idea if they speak or communicate in other human languages. Most use telepathy, communicating through the mind without the use of speech, signals, or body language. They also use imagery, graphics, and visions to present complicated concepts.

While some of the creatures I've encountered seem to understand human belief systems such as religion, morals and ideals, and the specific vocabulary we use to communicate ideas, not all understand us to this depth. I suspect some extraterrestrial races know very little about us and have no idea how to communicate.

For those who do, I believe the purpose of their communication is to trigger emotions, to connect, to make us feel. This reminds me of a rule of writing: show, don't tell. This technique allows the recipient to experience the story or idea through sensory details and emotions rather than exposition. In simple terms, showing illustrates, while telling merely states. And what better way to get your message across than to show your audience how something feels.

That said, there have been times I wished they would just spit it out—tell me what's going on straight up. They claim I wouldn't understand, but maybe they were wrong.

Their intentions are good.

I cannot tell you why creatures from other planets visit and abduct human beings. I am not omnipresent or all-knowing. I can't read minds or study extraterrestrials outside the scope of the experiences they wish to thrust upon me. My perspective is extremely limited.

What I do know is they are curious about us. I don't believe they see mankind as an inferior race or as livestock in need of corralling. We are currently no threat to them. I think they see us as beings who exist like they do, as equals who happen to live on a different planet or in another dimension. Perhaps they are even in awe of us.

I have no doubt they are here to learn, to see what we're made of and how we work. Some, like the Grays, seem calculated and focused on our physical nature, while others, like the aquatic beings, are in search of more. They want to know how we process information and not only survive but multiply and thrive. And yet, I don't believe they are all take and no give. They show an interest in our well-being, in our future as a species. They try to teach us, to share what they've learned.

They've seen things, been places, and have evolved in a way we humans could learn from—if we weren't so afraid. Maybe they could help us save our planet. There's a good chance they could prepare us for natural disasters—man-made and otherwise—and teach us how to respect all life, including our own. If we were willing to listen, connect, and feel, perhaps creatures from afar could offer humanity an altered fate.

Of one thing I am sure: they are willing to guide us, to advance our species—to be better human beings. Call it an awakening, spiritual Ascension, or whatever you wish. Either way, there is a natural evolutionary process, an inner expansion of the mind, heart, and soul, and every single one of us—perhaps every living thing—is on this path. While some people speak of spiritual Ascension in terms of being upgraded, rebooted, or elevated to a higher vibrational frequency, the meanings are all the same. Ascension is about tapping into our conscious minds, about change. It's transcending old, limiting habits, beliefs, and mindsets to be better people.

We don't all experience this change alone. Some of us are helped toward it by them.

One ascends by being loving, forgiving, and empathetic. Not just when it's convenient, but always. It takes effort, work, but compassion, respect, and integrity come from our core. Generosity and kindness can and should be our first reaction, our only action. Spiritual Ascension is not a religion or belief system. It is a choice, a way of being.

If this sounds too big, unrealistic, or downright silly, don't worry. The path is neither short nor quick, but you'll get there eventually. I believe you are where you are meant to be. I am, too. No one can tell you how to learn, not even aliens with prior experience. I certainly can't tell you. I have a hard time explaining it myself. I can only say it's a feeling, a gut knowing, a connection to something bigger than ourselves, something more. Everyone arrives at this destination—everyone ascends—from a different path. And no two souls are alike.

Either way, given the state of our world today, is anything more paramount than nurturing what we've got?

We must try.

There is more to our world than we know or understand.

I am pretty sure life is always several steps ahead of science. Perhaps this has never been more obvious than today. As I write this book, the entire world is scrambling to survive COVID-19. We think we have medicine, technology, and power—and we do—yet we are easily threatened by a microscopic virus we know very little about. What we mainly have is fear.

Life finds a way while science tries to keep up. There is so much we don't know.

Aliens exist, and they must live somewhere. Instead of considering ourselves almighty, maybe we should expand our reach to include other planets, galaxies, and dimensions we know nothing about, as well as the vast majority of our backyard—our oceans and forests—we've never explored. There is no excuse for our ignorance.

What could we learn if we looked at things differently? What if outside-the-box thinking became the norm? What if those with wild, unheard-of ideas were heard, taken seriously, and embraced? Would our understanding of our world change and grow?

Most of humanity's most prized accomplishments were birthed by those considered unorthodox, out-there, or nuts. Imagine if power, politics, religion, and social handcuffs didn't stand in the way of innovation. If people could face the unknown with an open mind, perhaps our collective experiences would fill the holes and give us the answers we thirst for.

What if we were the visitors versus the visited?

Every living thing exists on a different level of vibrational consciousness. Everything. Although we are not consciously aware of these vibrations, we feel them, sense them, and connect with them deep inside. Our vibrations are connected to our very nature—some might say our souls. I used to think the human form did not include a soul. To believe such a thing, one would have to consider mortality in a different light. I now believe otherwise.

We are not our bodies. Our bodies are merely vessels to experience the world around us, to give us senses to experience life. When our human form expires, our souls continue to learn, always striving for the next level of consciousness. Our souls are here, on earth, to learn, experience, and grow. This is not a quick process. It takes many, many lifetimes. Our souls have been climbing the Ascension ladder forever. Our souls are meant to achieve, over many lifetimes, an understanding of the true meaning of love, kindness, and caring, so we can apply what we've learned to our everyday experiences.

That sounds easy enough, but is it? We don't always learn the easy way.

Not only do we have souls, but they are possibly the most valuable thing we possess. Our soul is our sole reason for existing. Nurturing your soul should be your main purpose.

These are not concepts I was taught in my youth. They are not fantastical ideals garnered from religion or stories passed to me by previous generations. I believe they taught me this.

I don't have all the answers.

I have a lot of questions—an endless number of queries and wonderments. My mind is always swimming in possibilities.

How could it not, considering what I've been through?

Why me? I fear this question might plague me forever. Short of my blood type, I'm not special or unique in any way. Other than the extraordinary experiences noted within these pages, I have lived a rather ordinary life. Maybe I was in the wrong place at the wrong time. Perhaps I was in the right place at the right time. Was it something I said or did as a child? Is there something strange or unique about my DNA? Could my heart or its ability be something special?

I don't know. I suspect I might never know.

Do you know? I know you are out there, one of the thousands or millions who have experienced what I have. Some of you call yourselves abductees or experiencers. But most, possibly a massive number of people, wouldn't dare give themselves such a title. Maybe you are one of the many who can't even bring yourself to admit that what you've gone through is real.

I understand.

I know what it's like to be stolen from bed as a kid and have no one believe you. I know what it's like to be a young adult discovering life while dealing with abductions and the self-doubt and confusion that go with them—it can be too much to bear. I know what it's like to be a middle-aged man who can't share his deepest, darkest secrets with those around him, even those he holds dear, like his wife and kids. I know what it's like to worry about your career, your livelihood, and your credibility.

The risks seem insurmountable. The fear is debilitating.

I know what it's like to question yourself, to have moments you doubt your sanity.

I've felt what it's like to worry every waking moment. Will today be the day aliens don't bring me home? Will they hurt me? Will I survive?

You are not alone. I know, and I've felt the crippling grip of isolation.

Find your tribe. Connect with others who understand your plight. Get help—positive, supportive help. You are not in this alone. I'm sure there are many of us, and we need to be there for each other when the people around us can't be.

Imagine a world where we all found our voice.

My name is Robert Hunt.

This book is a record of my life, of sixty-three years as a human being who has seen and experienced the unexplained, the unfathomable, the impossible—and lived to talk about it.

I am not an authority on alien abduction, paranormal phenomena, or extraterrestrial events. I am not even a credible source on all abduction experiences. I am not a doctor who has interviewed or regressed hundreds of abduction victims. I am not a scientist making a case or argument for fact versus fiction. I can't speak for others, for those who have suffered more or other miseries at the hands of extraterrestrials. I am simply a man relaying my encounters with alien beings and how I question or interpret those experiences.

I have told you my story, my truth.

At times I've felt alone—very alone—yet I've been blessed with a loving wife, two beautiful children, family, friends, associates, and a large cast of inhuman teachers from afar. Having never experienced what I've seen, my family doesn't always understand, but they have my back nonetheless. The reality is, I have never been alone. And for that, I am eternally grateful.

I suppose I still worry about the naysayers and skeptics in some ways, but they don't scare me like they used to. Humans can do very little to hurt me these days. I'm older, wiser, and a lot less vulnerable. I no longer worry about supporting my family. My future looks more like retirement. I'm satisfied to live the time I have left surrounded by those on a path to greener pastures.

Still, I must confess. I worry about non-humans. In their hands, I am always the disadvantaged. While I have no doubt extraterrestrials have helped me on my journey, I don't know how they'll feel about this book. They've never suggested that what I was shown, taught, or experienced was a secret that couldn't be shared, but what if I'm wrong? What if they discover I've told all and are not happy about it?

Ah, another line of questions to add to my list.

Until my end, or one Sunday in August, I will walk this path we call life. I will welcome other-worldly visitors with open arms, despite my fears. I will be grateful for what I have, everything I've seen, and who I've become. And when my time comes, I will ascend to a vibrational level that feeds my soul with positive, compassionate, and loving goals.

Why me, you ask?

Because I am ready.

ACKNOWLEDGMENTS

I thank my wife, Lisa, for understanding why I needed to write about my alien encounters and Ascension. Lisa is genuinely uncomfortable with a controversial subject such as this.

Special thanks to Dee Willson, the accomplished writer who was willing to undertake such an awkward memoir and interject wisdom and experience to complete this book.

To you, the reader, the one willing to consider the nature of our reality by opening your mind to a truth right in front of you—I applaud you. Most people refuse to see.

RECOMMENDED READING

"The greatest mass of data comes not from physical findings but from the reports of the experiencers themselves. Although varied in some respects, these are so densely consistent as to defy conventional psychiatric explanations."

JOHN E. MACK, MD, *ABDUCTION*

People are coming to understand we are not alone in this universe. More is being shared about extraterrestrials and those they abduct or visit. At the time of writing this book, the following sources were a great place to start.

To be clear, the following research sources (books, organizations, websites, reports, articles, and live interviews) were found and studied after every one of my experiences were fully noted, discussed, written, and edited. Most of these sources were not known or reviewed by me before being shown me by my cowriter, Dee Willson, in an attempt to understand common threads between abductees.

In other words, the experiences of others had zero involvement or impact on the personal experiences noted within these pages. The events in this book are mine alone. Learning there are abductees from around the world—people who have never met or spoken to each other—who have shared the same or similar extraterrestrial experiences has been eye-opening, to say the least. Their stories have added to my puzzle pieces. Most of all, I was relieved to discover I was not alone in my experiences. This research offered me the sense of belonging I'd been deprived of for so long. And, for that, I am thankful.

Published Books

Abduction, John E. Mack, MD (Pulitzer Prize winner), Scribner (Macmillan Publishing Company), August 1, 2007.

The Day After Roswell, Col. Phillip J. Corso and William J. Birnes, first edition by Simon & Schuster, June 1, 1998; reissued by Gallery Books, June 6, 2017.

Communion, Whitley Strieber, Harper Collins/William Morrow Paperbacks, January 2, 2008.

The Grays, Whitley Strieber, first edition by Tor Books, August 22, 2006; reissued by Forge Books, May 29, 2007.

Unacknowledged, Stephen M. Greer, MD, Sirius Technology Advanced Research LLC, April 18, 2019.

Alien Abductions, Terry Matheson, Prometheus, November 1, 1998.

The Alien Abduction Files, Kathleen Marden and Denise Stoner, Career Press/New Page Books, May 20, 2013.

Into the Fringe, Dr. Karla Turner, Ph.D., Berkley Publishing, November 1, 1992.

Taken, Dr. Karla Turner, Ph.D., Kelt Works, April 1, 1994.

Secret Life, David M. Jacobs, Ph.D., first edition by Simon & Schuster, March 1, 1992; reissued by Touchstone, April 16, 1993.

Alien Implants, Dr. Roger Leir, Dell, August 8, 2000.

Extraterrestrial Contact, Kathleen Marden, Red Wheel Publishing, August 5, 2019.

Captured! The Betty and Barney Hill UFO Experience, Stanton T. Friedman and Kathleen Marden, first edition by New Age Books, July 23, 2007; reissued by New Age Books, April 1, 2021.

Project Blue Book, Brad Steiger and Donald R. Schmitt, first edition by

Ballantine Books, May 12, 1987; reissued by Red Wheel Publishing, February 1, 2019.

Over a dozen books written by Dolores Cannon, https://ozarkmt.com/ product-category/dolores-cannon/

Organizations and Websites

MUFON, www.MUFON.com

Association Québécoise d'Ufologie (AQU) (Canada) webmestre@ ovni-expert.com

GARPAN, https://garpan.ca/en/

NOUFORS, https://www.facebook.com/Northern-Ontario-UFO-Research-and-Study-NOUFORS

National UFO Reporting Center (NUFORC), http://www.nuforc.org/

UFO*BC, https://www.ufobc.ca/

UFO Info, http://ufoinfo.ca/

UFOROM, http://uforum.blogspot.com

Earthfiles, https://www.earthfiles.com/

The Black Vault, https://www.theblackvault.com/casefiles/

ICAR, http://www.ufoabduction.com/

Reports and Articles

The Marden-Stoner Study on Commonalities Among UFO Abduction Experiencers, https://www.kathleen-marden.com/commonalities-study-final-report.php

The Awakened Truth, www.theawakenedtruth.com

Unknown Country, www.unknowncountry.com

Live Interviews

The Joe Rogan Experience interview with Travis Walton, https://open.
spotify.com/episode/0mCfpeY0Ga4meTanFzOkkL

The Joe Rogan Experience interview with US Navy Commander David
Fravor, https://www.youtube.com/watch?v=Eco2s3-0zsQ

The Joe Rogan Experience interview with Bob Lazar, https://open.
spotify.com/episode/7Gg4Qi578G5SXoEtaLVVpx

The Joe Rogan Experience interviews (two) with Brian Cox, https://
open.spotify.com/episode/0lmJenV6c0Y55KcFZzRpdA
https://open.spotify.com/episode/10cY5SzMieX1FiW7Shd06Y

The Joe Rogan Experience interview with George Knapp and Jeremy
Corbell, https://open.spotify.com/episode/3RIsqi1Axn6zPGd0Ip
CRgf

The Joe Rogan Experience interviews (four) with Randall Carlson,
https://open.spotify.com/episode/3p6k8DKOKdlcg3yO5olZCy?
si=Iq2JqyDdTzGCBhY3vii_5w&nd=1
https://open.spotify.com/episode/5qKlK65lMbogGsJ9Aem0jo
https://open.spotify.com/episode/7s5FNLNjizgPAfNmfo03rt
https://open.spotify.com/episode/1baTI8QiylAMQ1zCf7HmIX

The Joe Rogan Experience interview with Christopher Mellon, https://
open.spotify.com/episode/2V0uWX1C4m8xEL0HHYqbnE

Government Authorities

NORAD (North American Aerospace Defense Command) confirmed
it responds to unknown, unwanted, and unauthorized air traffic
but not to reports of UFO sightings within the United States and
Canada, https://www.norad.mil/

Association Québécoise d'Ufologie (AQU) (Canada), webmestre@
ovni-expert.com

RECOMMENDED READING

National UFO Reporting Center (NUFORC), http://www.nuforc.
org/

Garpan, https://garpan.ca/en/tag/aqu/